Student Solutions Manual

to accompany

Analytical Chemistry

Sixth Edition

Gary D. Christian
University of Washington

WILEY

Gregory Manicol/Photo Researchers

To order books or for customer service call 1-800-CALL-WILEY (225-5945).

ISBN 0-471-26844-5

10 9 8 7 6 5 4

PREFACE

Solutions to all problems should be attempted before referring to the Solutions Manual. There is frequently more than one way to set up a problem, and your setup may differ from the one illustrated here. The detailed solutions given here are intended as a guide to help you get started when necessary.

Spreadsheet Problems

The solutions to all spreadsheet problems are given in the CD-ROM provided with the textbook.

CONTENTS

1. *The chemical characterization of matter.*
2. *Qualitative analysis deals with the identification of the presence of a particular substance or substances in a sample. Quantitative analysis deals with determining how much is present.*
3. *Define the problem, obtain a representative sample, dry the sample if required, measure its weight or volume, dissolve the sample and prepare the solution for the measurement step, measure the analyte, calculate the amount or concentration of analyte in the sample, and compute the precision of the analysis.*
4. *A sample represents the material to be analyzed. The analyte is the substance to be measured or determined. Hence, we determine the analyte by analyzing the sample.*
5. *A blank consists of all chemicals used in an analysis, run through the analytical procedure, to determine impurities that might be added to the analytical result, and which must be subtracted.*
6. *Gravimetry, volumetric analysis, instrumental analysis, kinetic methods of analysis, and combinations of these*
7. *Precipitation (gravimetry), chromatography, solvent extaction, volatilization (distillation)*
8. *The measurement of a physical property of the sample*
9. *A calibration curve represents an instrument (detector) response as a function of concentration. It may be a linear or a nonlinear response. An unknown analyte concentration in a sample solution is determined by comparison of the response with the calibration curve.*

10. *A specific reaction occurs only with the substance (analyte) of interest. Selective reaction occurs preferentially with the substance of interest, but not exclusively.*

11. *(a) Precipitate chloride with silver nitrate and weigh the purified precipitate. Measure sodium by atomic spectroscopy or ion-selective electrode to distinguish from KCl impurities. (b) Titrate with standard sodium hydroxide solution. (c) Measure potentiometrically with a pH meter/electrode.*

1. Volumetric flasks (t.c.), pipets (t.d.) (some micropipets t.c.), burets (t.d.).

2. It is a first class lever in which an unknown mass is balanced against a known mass. If each arm of the lever is equal in length, then the two masses at balance are equal.

3. Because the ratio of beam mass to length is decreased and the pan mass is decreased. The center of gravity is also adjusted for greater sensitivity.

4. The "TD" means "to deliver", and "TC" means "to contain" the specified volume.

5. The sample plus container is weighed, the sample is removed, and the loss in weight is the weight of the sample. This technique is useful for weighing hygroscopic samples that must be kept stoppered and for weighing several successive aliquots of the same sample.

6. The zero point is the equilibrium point of the balance under no load, while the rest point is the equilibrium point under load. In making a weighing, the rest point is made to coincide with the zero point.

7. Don't handle objects with the fingers, weigh objects at room temperature with the balance door closed, release the beam arrest (for a mechanical balance) and then secure the pan arrest (follow the reverse order in securing the arrests), secure the beam and pan arrests when adding or removing objects or weights, and never place chemicals directly on the pan.

8. Concentrated hydrochloric acid is diluted, preferably with boiled distilled water. It is standardized by titrating against primary standard sodium hydroxide or tris(hydroxymethyl)aminomethane. A saturated solution of sodium hydroxide is prepared and the insoluble sodium carbonate is allowed to settle out and then the supernatant is decanted. Or the saturated solution is filtered. The solution is diluted in boiled distilled water and standardized against primary standard potassium acid phthalate.

9. Dry ashing involves burning away the organic matter at an elevated temperature (400-700° C) with atmospheric oxygen as the oxidant. In wet digestion, the organic matter is oxidized to CO2, H2Op, and other products by a hot oxidizing acid. Dry ashing is relatively free from contamination, but it has the danger of loss by volatilization or rentention.

Wet digestion is relatively free from retention and volatility losses, but it has the danger of contamination from impurities in the reagents.

10. *Acid dissolution and acid or alkaline fusion followed by acid, neutral, or alkaline dissolution.*

11. *Protein free filtrate. It is prepared by mixing a biological fluid with a protein precipitating agent, such as trichloroacetic acid, tungstic acid, barium sulfate, etc., followed by filtering or centrifuging the precipitated proteins.*

12. *Care must be taken to prevent the digestion mixture from going too near dryness. Perchloric acid must not be added directly to organic or biological material, but only after an excess of nitric acid is added. The fumes from the digestion sho;uld be collected or else a specially designed hood used.*

13. *The gross sample is the entire collected sample that is representative of the whole. This is reduced to a size suitable for handling, called the sample. An aliquot of the sample, called the analytical sample, is weighed and analyzed. Several aliquots of the sample may be analyzed. A grab sample is a single random sample that is assumed to be representative of the whole, an assumption that is valid only for homogeneous samples.*

14. *The electric field of the microwave energy causes molecules with dipole moments to rotate to try to align with the electric field, and ions migrate in the electric field. These movements result in heat.*

15. *Weight in air of water contained:*

$$\begin{array}{r} 52.127\ g \\ \underline{-27.278\ g} \\ 24.849\ g \end{array}$$

$$W_{vac} = 24.849 + 24.849(0.0012/1.0 - 0.0012/7.8)$$

$$= 24.849 + 0.026 = 24.875\ g$$

$$V_{22^0} = 24.875\ g/0.99777\ g/mL = 24.931\ mL$$

$$V_{20^0} = 24.931\ mL\ (0.99777/0.99821) = 24.920\ mL.$$

16. *From Table 2.4:*

$$V_{25^0} = 24.971 \times 1.0040 = 25.071\ mL$$

$$V_{20^0} = 25.071\ mL \times (0.9970/0.9982) = 24.041\ mL.$$

17. *From Table 2.4, the volumes expand by the ratio of 1.0052/1.0028 = 1.0024. So (volumes all in mL):*

Nom. vol.	Vol. 20^0	Vol. 30^0	Change	Correction 30^0
10	10.02	10.04	+0.02	+0.04
20	20.03	20.08	+0.05	+0.08
30	30.00	30.07	+0.07	+0.07
40	39.96	40.06	+0.10	+0.06
50	49.98	50.10	+0.12	+0.10

18. *From Table 2.4:*

0.05129 x 0.9980/0.9962 = 0.05138 M.

The volume expansion is 0.18%, causing the concentration to decrease this amount.

CHAPTER 3

1. Accuracy is the agreement between the measured value and the accepted true value. Precision is the agreement between replicate measurements of the same quantity.

2. A determinate error is one that is non-random but is usually unidirectional and can be ascribed to a definite cause. An indeterminate error is a random error occurring by chance.

3. (a) determinate, methodic

   (b) determinate, methodic

   (c) indeterminate

   (d) determinate, instrumental

   (e) determinate, operative

4. (a) 5 (b) 4 (c) 3

5. (a) 4 (b) 4 (c) 5 (d) 3

6. Li (6.9417) + N (14.0067) + 3O (47.9982) = 68.9466

7. Pd (106.4) + 2Cl (70.9) = 177.3

8. $162._2$

9. $0.013_9 - 0.0067 + 0.00098 = 0.008_2$

10. To the nearest 0.01 g for three significant figures

11. (a) mean = 100 meq/L

    (b) absolute error = -2 meq/L

    (c) relative error = [-2/102] x 100% = -2%

12. (a) mean = 128.0 g

    (b) median = 128.1 g

    (c) range = 129.0 g - 127.1 g, or 1.9 g

13. (a) absolute error = 22.62 g - 22.57 g = 0.05 g

relative error = (0.05 g/22.57 g) x 100% = 0.2_2% = $2._2$ ppt

(b) absolute error = 45.02 mL - 45.31 mL = -0.29 mL

relative error = (-0.29 mL/45.31 mL) x 100% = -0.64% = -6.4 ppt

(c) absolute error = 2.68% - 2.71% = -0.03%

relative error = [(-0.03%)/(2.71%)] x 100% = $-1._1$% = -1_1 ppt

(d) absolute error = 85.6 cm - 85.0 cm = 0.6 cm

relative error = [(0.6 cm)/(85.0 cm)] x 100% = 0.7% = 7 ppt

14. (a) mean = 33.33% (See CD for spreadsheet calculations)

x_i	$x_i-\bar{x}$	$(x_i-\bar{x})^2$
33.27	0.06	0.0036
33.37	0.04	0.0016
33.34	0.01	0.0001
		Σ 0.0053

$s = \sqrt{(0.0053)/(3-1)}$ = 0.052% (absolute)

coeff. of varn = % rel. s = (0.052/33.33) x 100% = 0.16% (relative)

(b) mean = 0.024%

x_i	$x_i-\bar{x}$	$(x_i-\bar{x})^2$
0.022	0.002	4×10^{-6}
0.025	0.001	1×10^{-6}
0.026	0.002	4×10^{-6}
		Σ 9×10^{-6}

$s = \sqrt{(9 \times 10^{-6})/(3-1)}$ = 0.0021% (absolute)

coeff. of varn = (0.0021/0.024) x 100% = 8.8% (relative)

15. mean = 102.6 (See CD for spreadsheet calculations)

x_i	$x_i - \bar{x}$	$(x_i - \bar{x})^2$
102.2	0.4	0.16
102.8	0.2	0.04
103.1	0.5	0.25
102.3	0.3	0.09
		Σ 0.54

(a) $s = \sqrt{(0.54)/(4-1)} = 0.42$ ppm

(b) rel. s = (0.42/102.6) x 100% = 0.41%

(c) $s_{(mean)} = 0.42/\sqrt{4} = 0.21$ ppm

(d) rel. $s_{(mean)}$ = (0.21/102.6) x 100% = 0.20%

16. mean = 95.65% (See CD for spreadsheet calculations)

x_i	$x_i - \bar{x}$	$(x_i - \bar{x})^2$
95.67	0.02	0.0004
95.61	0.04	0.0016
95.71	0.06	0.0036
95.60	0.05	0.0025
		Σ 0.0081

(a) $s = \sqrt{(0.0081)/(4-1)} = 0.052\%$ (absolute)

(b) $s_{(mean)} = (0.052)/\sqrt{4} = 0.026\%$ (absolute)

(c) rel. $s_{(mean)}$ = [0.026/95.65] x 100% = 0.027% (relative)

17. (a) $s_a = [(\pm 2)^2 + (\pm 8)^2 + (\pm 4)^2]^{\frac{1}{2}} = (\pm 84)^{\frac{1}{2}} = \pm 9.1$

128 + 1025 - 636 = 517 ± 9

(b) $s_a = [(\pm 0.06)^2 + (\pm 0.03)^2]^{\frac{1}{2}} = (\pm 0.0045)^{\frac{1}{2}} = \pm 0.067$

16.25 - 9.43 = 6.82 ± 0.07

(c) $s_a = [(\pm 0.4)^2 + (\pm 1)^2]^{\frac{1}{2}} = (\pm 1.16)^{\frac{1}{2}} = \pm 1.1$

46.1 + 935 = 981 ± 1

(See CD for spreadsheet calculation of s_a)

18. (a) $(2.78 \pm 0.04)(0.00506 \pm 0.00006) = 0.01407 \pm ?$

$(s_b)_{rel} = (\pm 0.04)/(2.78) = \pm 0.014$

$(s_c)_{rel} = (\pm 0.00006)/(0.00506) = \pm 0.012$

$(s_a)_{rel} = [(\pm 0.014)^2 + (\pm 0.012)^2]^{\frac{1}{2}} = (\pm 0.00034)^{\frac{1}{2}} = \pm 0.018$

$s_a = (0.01407)(\pm 0.018) = \pm 0.0003$ *(See CD for spreadsheet calculation of* $(s_a)_{rel}$*)*

∴ the answer is *0.0141 ±0.0003*

(b) $(36.2 \pm 0.4)/(27.1 \pm 0.6) = 1.336 \pm ?$

$(s_b)_{rel} = (\pm 0.4)/(36.2) = \pm 0.011$

$(s_c)_{rel} = (\pm 0.6)/(27.1) = \pm 0.022$

$(s_a)_{rel} = [(\pm 0.011)^2 + (\pm 0.022)^2]^{\frac{1}{2}} = (\pm 0.00060)^{\frac{1}{2}} = \pm 0.024$

$s_a = (1.336)(\pm 0.024) = \pm 0.032$

∴ the answer is 1.34 ±0.03

(c) $(50.23 \pm 0.07)(27.86 \pm 0.05)/(0.1167 \pm 0.0003) = 11{,}991 \pm ?$

$(s_b)_{rel} = (\pm 0.07)/(50.23) = \pm 0.0014$

$(s_c)_{rel} = (\pm 0.05)/(27.86) = \pm 0.0018$

$(s_d)_{rel} = (\pm 0.0003/0.1167) = \pm 0.0026$

$(s_a)_{rel} = [(\pm 0.0014)^2 + (\pm 0.0018)^2 + (\pm 0.0026)^2]^{\frac{1}{2}} = \pm 0.0035$

$s_a = (11{,}991)(\pm 0.0035) = \pm 42$

∴ the answer is 11,990±40 $[(1.199 \pm 0.004) \times 10^4]$

19. $[(25.0 \times 0.0215 - 1.02 \times 0.112)(17.0)]/(5.87)$

$= (0.538 - 0.114)(17.0)/(5.87) = (0.424)(17.0)/(5.87) = 1.228 \pm ?$

For 25.0×0.0215:

$(s_b)_{rel} = (\pm 0.1)/(25.0) = \pm 0.0040$

$(s_c)_{rel} = (\pm 0.0003)/(0.0215) = \pm 0.014$

$(s_a)_{rel} = [(\pm 0.0040)^2 + (\pm 0.014)^2]^{\frac{1}{2}} = \pm 0.015$

$s_a = (0.538)(\pm 0.015) = \pm 0.0081;\ \therefore\ 0.538 \pm 0.008$

For 1.02×0.112:

$(s_b)_{rel} = (\pm 0.01)/(1.02) = \pm 0.0098$

$(s_c)_{rel} = (\pm 0.001)/(0.112) = \pm 0.0089$

$(s_a)_{rel} = [(\pm 0.0098)^2 + (\pm 0.0089)^2]^{\frac{1}{2}} = \pm 0.013$

$s_a = (0.114)(\pm 0.013) = \pm 0.0015;\ \therefore\ 0.114 \pm 0.002$

For $(0.538 \pm 0.008) - (0.114 \pm 0.002)$:

$s_a = [(\pm 0.0081)^2 + (\pm 0.0015)^2]^{\frac{1}{2}} = \pm 0.0082;\ \therefore 0.423 \pm 0.008$

For $[(0.423 \pm 0.008)(17.0 \pm 0.2)]/(5.87 \pm 0.01)$:

$(s_b)_{rel} = (\pm 0.0082)/(0.423) = \pm 0.019$

$(s_c)_{rel} = (\pm 0.2)/(17.0) = \pm 0.012$

$(s_d)_{rel} = (\pm 0.01)/(5.87) = \pm 0.0017$

$(s_a)_{rel} = [(\pm 0.019)^2 + (\pm 0.012)^2 + (\pm 0.0017)^2]^{\frac{1}{2}} = \pm 0.023$

$s_a = (1.225)(\pm 0.023) = \pm 0.028$

Answer = 1.22 ± 0.03

20. mean = 0.5027 $\underline{M}$ *(See CD for spreadsheet calculation of s)*

x_i	$x_i-\bar{x}$	$(x_i-\bar{x})^2$
0.5026	0.0001	1×10^{-8}
0.5029	0.0002	4×10^{-8}
0.5023	0.0004	16×10^{-8}
0.5031	0.0004	16×10^{-8}
0.5025	0.0002	4×10^{-8}
0.5032	0.0005	25×10^{-8}
0.5027	0.0000	0
0.5026	0.0001	1×10^{-8}
		$\Sigma\ 67 \times 10^{-8}$

$s = \sqrt{(67 \times 10^{-8})/(8-1)} = 0.00031$ M

From Equation 3.9 and Table 3.1 ($t = 2.365$ for $\nu = 7$ at 95% C.L.):

Conf. limit $= 0.5027 \pm (2.365 \times 0.00031)/\sqrt{8}$

$= 0.5027 \pm 0.00026$ M or $0.5024 - 0.5030$ M

21. mean = 139.6 meq/L (See CD for spreadsheet calculation of s)

x_i	$x_i-\bar{x}$	$(x_i-\bar{x})^2$
139.2	0.4	0.16
139.8	0.2	0.04
140.1	0.5	0.25
139.4	0.2	0.04
		$\Sigma\ 0.49$

$s = \sqrt{0.49/(4-1)} = 0.40$ meq/L

(a) For 3 degrees of freedom at the 90% confidence limit, $t = 2.353$

Conf. limit $= 139.6 \pm (2.353 \times 0.40)/\sqrt{4}$

$= 139.6 \pm 0.47$ meq/L or 139.1 - 140.1 meq/L

(b) $t = 3.182$ at 95% confidence level

Conf. limit $= 139.6 \pm (3.182 \times 0.40)/\sqrt{4}$

$= 139.6 \pm 0.64$ meq/L or 139.0 - 140.2 meq/L

(c) $t = 5.841$ at 99% confidence level

Conf. limit $= 139.6 \pm (5.841 \times 0.40)/\sqrt{4}$

$= 139.6 \pm 1.17$ meq/L or 138.4 - 140.8 meq/L

Note that in order to be more confident of the range of the true value, the range must increase. Conversely, as the range is narrowed, we are less confident that it defines the true value.

22. t for 2 degrees of freedom at 90% confidence level = 2.920

Conf. limit $= \pm(2.920 \times 2.3)/\sqrt{3} = \pm 3.9$ ppm

23. t for 2 degrees of freedom at 95% confidence level = 4.303

Conf. limit $= \pm(4.303 \times 0.5)/\sqrt{3} = \pm 1._2$ meq/L

24. Applying the Q-test to the standardization data shows that 0.1050 should probably be rejected. Then the valid data are 0.1071, 0.1067 and 0.1066. The standard deviation for these triplicate results calculates to be 0.00026 M (mean 0.1068 M). t = 2.920

Conf. limit $= 0.1068 \pm (2.920 \times 0.00026)/\sqrt{3}$

$= 0.1068 \pm 0.0004$ M or 0.1064 - 0.1072 M

25. $\bar{x}_C = 15._8$; $\bar{x}_D = 23._3$

$(x_i - \bar{x}_C)^2$	$(x_i - \bar{x}_D)^2$
1	4
49	9
16	144
4	81
49	64
144	289

25	49
36	169
	1
	25
Σ 32_4	Σ 83_5

F-test: $F = (s_D^2)/(s_C^2) = (83_5/9)/(32_4/7) = 2.0_0$ *(See CD)*

$F_{Table} = 3.68$. Therefore, the precision of the two groups is comparable and the t-test can be applied. Use the paired t-test.

$s_\mu = \sqrt{(32_4 + 83_5)/(8 + 10 - 2)} = 8.51$

$\pm t = (23._3 - 15._8)/8.51 \sqrt{(10 \times 8)/(10 + 8)} = 1.8_6$

This is smaller than the tabulated t value for 16 degrees of freedom at the 95% confidence level, but not at the 90% confidence level. It appears there is a fair probability the differences between the two populations is real. More studies are indicated.

26. $\bar{x}_E = 13.0_0$; $\bar{x}_G = 13.1_8$

$(x_i-\bar{x}_E)^2$	$(x_i-\bar{x}_G)^2$
0.01	0.09
0.09	0.01
0.16	0.04
0.09	0.09
0.09	Σ 0.23
Σ 0.44	

First perform an F-test.

$F = (s_E^2)/(s_G^2) = (0.44/4)/(0.23/3) = 1.4_3$ *(See CD)*

$F_{Table} = 9.12$. Hence, there is a high probability the variances of the two methods represent the same population variance (note the standard deviations are nearly identical).

Apply the paired t-test.

$s_\mu = \sqrt{(0.44 + 0.23)/(9 - 2)} = 0.31$

$\pm t = (13.0_0 - 13.1_8)/0.31 \sqrt{(5 \times 4)/(5 + 4)} = 0.8_7$

t_{Table} = 2.365, so there is a high probability the two methods give the same result.

27.

$(x_i-\bar{x}_A)^2$	$(x_i-\bar{x}_B)^2$	$(x_i-\bar{x}_C)^2$
0.00036	0.00026	0.00102
0.00122	0.00012	0.00029
0.00122	0.00002	0.00040
0.00040	Σ 0.00040	0.00002
Σ 0.00320		Σ 0.00173

(See CD for spreadsheet calculation)

$s_p = \sqrt{(0.00320 + 0.00040 + 0.00173)/(11 - 3)} = 0.0258$ absorbance units

28. Colorimetry: $s_1^2 = [\Sigma(D_i-\bar{D})^2]/(N - 1) = 6.53/(8 - 1) = 0.933$

AAS: $s_2^2 = (1.67)/(6 - 1) = 0.334$

$F = (s_1^2)/(s_2^2) = (0.933)/(0.334) = 2.79$ (See CD)

F_{Table} for $\nu_1 = 7$ and $\nu_2 = 5$ is 4.88. Since $F_{calc} < F_{Table}$, there is no significant difference in the two variances.

29. The mean is 0.1017 *M* and the standard deviation is 0.0001_7 *M*.

$\pm t = (\bar{x}-\mu)(\sqrt{N}/s) = (0.1017 - 0.1012)(\sqrt{4}/0.0001_7) = 5._9$ (See CD)

This exceeds the tabulated t value even at the 99% confidence level, so there is a 99% probability that the difference is real and not due to chance.

30. The mean is 99.89% with a standard deviation of 0.033%.

$\pm t = (\bar{x}-\mu)(\sqrt{N}/s) = (99.89 - 99.95)(\sqrt{4}/0.033) = 3._6$ (See CD)

This just exceeds the tabulated t value at the 95% confidence level. Hence, there is a 95% probability that the analyzed data are significantly different from the supplier's stated value. Note that the difference of 0.060% is about twice the standard deviation, and we would expect this to occur by chance only 1 out of 20 (5%). Whether the shipment is accepted depends on the acceptable differences.

31. Arrange in decreasing order:

0.1071, 0.1067, 0.1066, 0.1050

The suspect result is 0.1050.

$Q = (0.0016)/(0.0021) = 0.76$

Tablulated Q = 0.829. Hence, it is 95% certain that the suspected value is not due to accidental error.

32. *For the Zn determination:*

33.37, 33.34, 33.27

The suspect result is 33.27.

$Q = (0.07)/(0.10) = 0.70$

Q_{Table} = 0.970 *Therefore, the number 33.27 is valid.*

For the Sn determination:

0.026, 0.025, 0.022

The suspect result is 0.022

$Q = (0.003)/(0.004) = 0.75$

Q_{Table} = 0.970 *Therefore, 0.022 is a valid result.*

33. *Arranging in order:*

22.25, 22.23, 22.18, 22.17, 22.09

$Q = (0.08)/(0.16) = 0.50$

Q_{Table} = 0.710. *Therefore, 22.09 is a valid measurement.*

34. *The range is 103.1 - 102.2 = 0.9 ppm*

From Equation 3.17 and Table 3.4 for 4 observations,

$s_r = (0.9)(0.49) = 0.44$ *ppm.*

This compares with s = 0.42 ppm calculated in Problem 15.

35. The range is 0.5032 - 0.5023 = 0.0009 M.

From Equation 2.18 and Table 2.4 for 8 observations,

conf. limit = 0.5027 ± 0.0009(0.29) = 0.5027 ± 0.00026

or 0.5024 - 0.5030.

This is identical to the confidence limit calculated using the standard deviation.

36. The range is 140.1 - 139.2 = 0.9 meq/L.

From Equation 3.18 and Table 3.4 for 4 observations,

$\text{conf. limit}_{95\%}$ = 139.6 ± 0.9(0.72) = 139.6 ± 0.65

or 139.0 - 140.2, the same as using the standard deviation.

$\text{conf. limit}_{99\%}$ = 139.6 ± 0.9(1.32) = 139.6 ± 1.19

or 138.4 - 140.8, the same as using the standard deviation.

37.

$(x_i-\bar{x})$	$(x_i-\bar{x})^2$	$(y_i-\bar{y})$	$(x_i-\bar{x})(y_i-\bar{y})$
-0.300	0.0900	-16.7	5.01
-0.200	0.0400	-10.9	2.18
-0.100	0.0100	- 4.5	0.45
0.100	0.0100	5.6	0.56
0.500	0.2500	26.6	13.3_0
	Σ 0.4000		Σ 21.5_0

$m = (21.50)/(0.4000) = 53.7_5$ (See CD)

This is identical to the value obtained with Equation 3.23. See Example 3.21.

38.

x_i	y_i	x_i^2	$x_i y_i$
1.00	0.205	1.00	0.205
2.00	0.410	4.00	0.820
3.00	0.615	9.00	1.84_5
4.00	0.820	16.00	3.28_0
Σ 10.00	Σ 2.050	Σ 30.00	Σ 6.15_0

$(\Sigma x_i)^2 = 100.0$

$\bar{x} = (\Sigma x_i)/n = 2.500$ $\bar{y} = (\Sigma y_i)/n = 0.5125$ $n = 4$

Using Equations 3.23 and 3.22:

$m = [6.15_0 - (10.00 \times 2.050)/4]/(30.0_0 - 100.0/4) = 0.205$

$b = 0.5125 - (0.205)(2.500) = 0.000$

$y = 0.205x + 0.000$

Unknown:

$0.625 = 0.205x + 0.000$

$x = 3.05$ ppm P in urine

(See CD for spreadsheet for plot and calculation)

39. From Problem 38

$\Sigma y_i^2 = (0.205)^2 + (0.410)^2 + (0.615)^2 + (0.820)^2 = 1.260_7$

$(\Sigma y_i)^2 = (2.050)^2 = 4.202$

$\Sigma x_i^2 = 30.00$; $(\Sigma x_i)^2 = 100.0$; $m^2 = (0.205)^2 = 0.0420$

From Equation 3.24,

$s_y = \sqrt{[(1.260_7 - 4.202/4) - 0.0420\,(30.00 - 100.0/4)]/(4 - 2)}$

$= \pm 0.01_0$ absorbance.

From Equation 3.25,

$s_m = \sqrt{(0.01_0)/(30.00 - 100.00/4)} = \pm 0.004_5$ absorbance/ppm

$m = 0.205 \pm 0.004$

From Equation 3.26,

$s_b = 0.01_0\sqrt{30.00/[4\,(30.00) - 100.00]} = \pm 0.01_2$ absorbance

$b = 0.00_0 \pm 0.01_2$

The phosphorus concentration in the urine sample is given by

$x = [(0.62_5 \pm 0.01_0) - (0.00_0 \pm 0.00_1)]/(0.205 \pm 0.004) = 3.05 \pm ?$

$s_{num} = \sqrt{(\pm 0.01_0)^2 + (\pm 0.00_1)^2} = \pm 0.01_0$

$(s_{div})_{rel} = \sqrt{\pm(0.01_0/0.62_5)^2 + (\pm 0.004/0.205)^2} = \pm 0.02_5$

$s_{div} = 3.05\,(\pm 0.02_5) = \pm 0.07_6$

$x = 3.05 \pm 0.08$ ppm P in urine

40.

% yeast extract (x_i)	Toxin, mg (y_i)
1.000	0.487
0.200	0.260
0.100	0.195
0.010	0.007
0.001	0.002
Σ 1.311	Σ 0.951

$\bar{x}_i = 1.311/5 = 0262$ $\qquad$ $\bar{y}_i = 0.951/5 = 0.190$

$\Sigma x_i^2 = 1.050$ $\qquad$ $\Sigma y_i^2 = 0.343$

$n = 5$ $\qquad$ $\Sigma x_i y_i = 0.559$

$r = (n\Sigma x_i y_i - \Sigma x_i \Sigma y_i)/\{[n\Sigma x_i^2 - (\Sigma x_i)^2][n\Sigma y_i^2 - (\Sigma y_i)^2]\}^{\frac{1}{2}}$

$= (2.795 - 1.247)/[(5.250 - 1.719)(1.715 - 0.904)]^{\frac{1}{2}} = 0.915$ *(See CD)*

$r^2 = 0.84$

.˙. There is a good correlation between yeast extract concentration and the amount of toxin produced.

41.

Toxin, mg (x_i)	Dry weight (y_i)
0.487	116
0.260	53
0.195	37
0.007	8
0.002	1
0.951 = Σx_i	215 = Σy_i

$\bar{x}_i = 0.951/5 = 0.190$ $\bar{y}_i = 215/5 = 43$

$\Sigma x_i^2 = 0.343$ $\Sigma y_i^2 = 17699$

$\Sigma x_i y_i = 77.5$

$r = (387.5 - 204.5)/\sqrt{(1.715 - 0.904)(88495 - 46225)} =$

$(183.0)/\sqrt{(0.811)(42270)} = (183.0)/(185) = 0.989$; $r^2 = 0.978$ *(See CD)*

.˙. There is a strong correlation between fungal dry weight and amount of toxin produced.

42. *(See CD for spreadsheet calculations)*

Enzyme method (x_i)	Colori-metric method (y_i)	D_i	D_i-D	$(D_i\text{-}D)^2$	$x_i y_i$	x_i^2	y_i^2
305	300	5	1	1	91,500	93,025	90,000
385	392	- 7	-11	121	150,920	148,225	153,664
193	185	8	4	16	35,705	37,249	34,225
162	152	10	6	36	24,624	26,244	23,104
478	480	- 2	- 6	36	229,440	228,484	230,400
455	461	- 6	-10	100	209,755	207,025	212,521
238	232	6	2	4	55,216	56,644	53,824
298	290	8	4	16	86,420	88,804	84,100
408	401	7	3	9	163,608	166,464	160,801
323	315	8	4	16	101,745	104,329	99,225
Σ3,245	Σ3,208	Σ 37		Σ 355	Σ1,148,933	Σ1,156,493	Σ1,141,864

$D = 3.7$

From Equation 3.16 :

$s_d = \sqrt{(355)/(10 - 1)} = 6.28$

From Equation 2.15 :

$t = [3.7/6.28]\sqrt{10} = 1.8_6$

From Table 2.1 at the 95% confidence level and $\nu = 9$, $t = 2.262$. Since $t_{calc} < t_{table}$, there is no significant difference between the methods at the 95% confidence level.

From Equation 3.28., we calculate:

$r = [10(1,148,933)-(3,245)(3,208.)]/$

$\{[10(1,156,493)-(3,245)^2][10(1,141,864)-(3,208)^2]\}^{\frac{1}{2}} = 0.999$; $r^2 = 0.998$

Hence, there is a high degree of correlation.

43. *The average blank reading is 0.18, and the standard deviation is ±0.06.*

The net reading for the detection limit is 3 x 0.06 = 0.18.

The net reading for the standard is 1.25 - 0.18 = 1.07.

The detection limit is 1.0 ppm (0.18/1.07) = 0.17 ppm.

This would give a total (blank plus analyte) reading of 0.18 + 0.18 = 0.36.

44. $WR^2 = K_s$

$K_s = (0.4\ g)\ (5)^2 = 10\ g$

For $R = 2.5$,

$W(2.5)^2 = 10$ g

$W = 1.6$ g. sample

45. $s_s = 0.15\%$ (wt/wt)

$R = 0.05$ and $s_x = (0.05)(3\%) = 0.15\%$ (wt/wt)

$x = 3\%$ (wt/wt)

From Equation 3.31

$n = t^2 s_s^2 / R^2 \bar{x}^2 = (1.96)^2 (0.15)^2 / (0.05)^2 (3)^2 = 3.84$

or from Equation 3.32

$n = t^2 s_s^2 / s_x^2 = (1.96)^2 (0.15)^2 / (0.15)^2 = 3.84$

For $n = 4$, $t = 2.776$:

$n = (2.776)^2 (1) = 7.71$

n is between 8 and 4, try 6. For $n = 6$, $t = 2.447$

$n = (2.447)^2 (1) = 5.99 = 6$ analyses required.

CHAPTER 4

1. Good Laboratory Practice is the general principle of assuring quality operation of a laboratory, from management practices, laboratory personnel, method validation and quality control, to reporting and record keeping, in order to assure correctness of results produced by the laboratory.

2. QUA is the Quality Assurance Unit, who is responsible for implementing and assessing quality procedures, and Standard Operating Procedures (SOPs) that provide details for carrying out the laboratory operations.

3. See Question 2 above.

4. The QUA should be independent from the laboratory. It establishes the quality assurance and quality control procedures to be implemented, and monitors and assesses them.

5. The problem is first defined, along with the data requirements. Then performance of the selected method must be validated to meet these requirements.

6. The minimum requirements of the method are decided, including accuracy and precision.

7. A technique refers to the technology to be used for a measurement, e.g., spectrophotometry. A method is the application of the technique, developing the proper chemistry or procedure for selective measurement. A procedure is the written directions for using the method. A protocol is a set of specifically prescribed directions that must be followed for official acceptance of the results.

8. Method validation generally requires studies to determine and validate selectivity, linearity, accuracy, precision, sensitivity, range, limit of detection, limit of quantitation, and ruggedness or robustness.

9. The Response Factor is a way of assessing linearity, by determining if the response (with y-intercept subtracted) per unit concentration remains reasonably constant over the concentration ranges.

10. Besides the Response Factor (Question 9), the coefficient of determination (r^2), and a small y-intercept are measures of linearity range.

11. Accuracy is determined by recovery studies, comparing results with those of another method of known accuracy, or by analyzing a reference material. The ultimate measure of accuracy is from analysis of a standard reference material.

12. At least seven measurements (six degrees of freedom) should be made.

13. Repeatability = short term intralaboratory precision.
Ruggedness = long term intralaboratory precision.
Robustness (repeatability) = sensitivity to small changes in parameters.
Reproducibility (transferability) = interlaboratory precision or bias.

14. Eelctronic records need to be backed up, archived, and recoverable. The data have to be secure, and any changes documented, with retention of original data. They must be transferable if the software is changed. There must be time- and date-stamped audit trails that can't be changed. Electronic signatures require both a username and password that are unique, not reassignable. The password should be changed periodically.

15. QA is the ongoing checking of the performance of a method. It includes appropriate quality control procedures, which provide quantitative measures of performance.

16. Quality control activities include maintaining control charts, using blind and random reference samples, and proficiency testing via collaborative laboratory studies.

CHAPTER 4

17. A z-score is a measure of how close a laboratory's result in a collaborative study are to the known concentration, by how many standard deviations of the accepted concentration it differs.

18. Accreditation is when an authoritative body certifies a laboratory is competent to perform specified tasks.

19. See CD for spreadsheets.

Least squares plot: $y = 2050x - 2$

Response Factor:

$RF = (y - 2)/x$

Ave. $RF = 1999$

Slope = 271RF/1% BA

RF change 0.02 to 0.20% = 271x0.18 = 49

% change = (49/19099) x 100% = 2.4%

20. A. $C = 1\ mg/kg = 10^{-6}$

$S_R = 0.02C^{0.85}$

$= 0.02(10^{-6})^{0.85}$

Spreadsheet cell: =.02(10^-6)^.85

= 1.55886E-07

$S_R = 1.6x10^{-7}$

$\%rsd = (1.6x10^{-7})/(10^{-6})x100 = 16\%$

B. $\%rsd = 2C^{-0.15}$

Spreadsheet cell: =2(10^-6)^-.15

= 15.88656469 = 16%

21. $z = (9.8 - 10.3)/0.5 = -1.0$. *Your results are within one standard deviation of the accepted value.*

1. *The first unit in each sample denotes the analyte unit, and the second denotes the sample unit. So the volume or the weight of either or both may be measured.*

2. *ppm* $= \mu g/g$ or mg/kg (wt/wt) $= g/g \times 10^6$

   $= \mu g/mL$ or mg/L (wt/vol) $= g/mL \times 10^6$

   $= nL/mL$ or $\mu L/L$ (vol/vol) $= mL/mL \times 10^6$

   ppb $= ng/g$ or $\mu g/kg$ (wt/wt) $= g/g \times 10^9$

   $= ng/mL$ or $\mu g/L$ (wt/vol) $= g/mL \times 10^9$

   $= pL/mL$ or nL/L (vol/vol) $= mL/mL \times 10^9$

3. *Eq. wt. = f.w./charge. This concept is used by physicians to give an overall view of the electrolyte balance.*

4. *A titration reaction should be stoichiometric and rapid, specific with no side reactions, quantitative, and there should be a marked change in a property of the solution at the equivalence point (when the reaction is complete). The four classes of titration are: Acid-base, reduction-oxidation, precipitation, and complexometric.*

5. *The equivalence point of a titration is the point at which the reaction is complete, and the end point is the point at which it is observed to be complete.*

6. *A standard solution is one whose concentration is known to the degree of accuracy required in an analysis (e.g., titration). It is prepared by dissolving a known amount of sufficiently pure reagent (e.g., a primary standard) in a known volume of solvent, or else by titrating a known quantity of a pure reagent (primary standard) with an approximately prepared solution to standardize it.*

7. *A primary standard should be* $\geq 99.98\%$ *pure, be stable to drying temperatures, have a high formula weight, and possess the properties required for a titration.*

8. *So that a sufficiently large amount of it will have to be weighed for the titration that the error in weighing is small.*

CHAPTER 5

9. (a) *5.00 g/100 mL x (250)/(100) = 12.5 g*

   (b) *1.00 g/100 mL x (500)/(100) = 5.00 g*

   (c) *10.0 g/100 mL x (1000)/(100) = 100 g*

10. (a) *[52.3 g/1000 mL] x 100 mL = 5.23 g/100 mL = 5.23% (wt/vol)*

   (b) *[275 g/500 mL] x 100 mL = 55.0 g/100 mL = 55.0% (wt/vol)*

   (c) *[3.65 g/200 mL] x 100 mL = 1.82 g/100 mL = 1.82% (wt/vol)*

11\. (a) 244.27 (b) 218.16 (c) 431.73 (d) 310.18

12\. (a) 500 mg/253 mg/mmol = 1.98 mmol $BaCrO_4$

(b) 500 mg/119 mg/mmol = 4.20 mmol $CHCl_3$

(c) 500 mg/389 mg/mmol = 1.28 mmol $KIO_3.HIO_3$

(d) 500 mg/137 mg/mmol = 3.65 mmol $MgNH_4PO_4$

(e) 500 mg/223 mg/mmol = 2.24 mmol $Mg_2P_2O_7$

(f) 500 mg/382 mg/mmol = 1.31 mmol $FeSO_4.C_2H_4(NH_3)_2SO_4.4H_2O$

13\. 0.200 mol/L x 0.100 L = 0.0200 mol of each substance required

(a) 253 g/mol x 0.0200 mol = 5.06 g $BaCrO_4$

(b) 119 g/mol x 0.0200 mol = 2.38 g $CHCl_3$

(c) 390 g/mol x 0.0200 mol = 7.80 g $KIO_3.HIO_3$

(d) 137 g/mol x 0.0200 mol = 2.74 g $MgNH_4PO_4$

(e) 223 g/mol x 0.0200 mol = 4.46 g $Mg_2P_2O_7$

(f) 382 g/mol x 0.0200 mol = 7.64 g $FeSO_4.C_2H_4(NH_3)SO_4.4H_2O$

14\. (a) mg NaCl = 1.00 mmol/mL x 1000 mL x 58.4 mg/mmol = 5.84×10^4 mg

(b) mg sucrose = 0.200 mmol/mL x 500 mL x 342 mg/mmol = 3.42×10^4 mg

(c) mg sucrose = 0.500 mmol/mL x 10.0 mL x 342 mg/mmol = 1.71×10^3 mg

(d) mg Na_2SO_4 = 0.200 mmol/mL x 10.0 mL x 142 mg/mmol = 284 mg

(e) mg KOH = 0.500 mmol/mL x 250 mL x 56.1 mg/mmol = 7.01×10^3 mg

(f) mg NaCl = 0.900 g/100 mL x 250 mL x 1000 mg/g = 2.25×10^3 mg

15\. (a) mL_{HCl} = 50.0 mmol/(0.100 mmol/mL) = 500 mL

(b) mL_{NaOH} = 10.0 mmol/(0.0200 mmol/mL) = 500 mL

(c) mL_{KOH} = 100 mmol/(0.0500 mmol/mL) = 2.00×10^3 mL

(d) $mL_{HBr} = 5.00\ g/(10.0\ g/100\ mL) = 50.0\ mL$

(e) $mL_{Na_2CO_3} = 4.00\ g/(5.00\ g/100\ mL) = 80.0\ mL$

(f) $1.00\ mol\ HBr = 80.9\ g$

$mL_{HBr} = 80.9\ g/(10.0\ g/100\ mL) = 809\ mL$

(g) $0.500\ mol\ Na_2CO_3 = 0.500 \times 106.0\ g/mol = 53.0\ g$

$mL_{Na_2CO_3} = 53.0\ g/(5.00\ g/100\ mL) = 1.06 \times 10^3\ mL$

16. $mmol\ Mn^{2+} = mmol\ Mn(NO_3)_2 = 0.100\ mmol/mL \times 10.0\ mL = 1.00\ mmol$

$\underline{M}_{Mn^{2+}} + 1.00\ mmol/30.0\ mL = 0.0333\ mmol/mL$

$mmol\ NO_3^- = mmol\ KNO_3 + 2 \times mmol\ Mn(NO_3)_2$

$= 0.100\ mmol/mL \times 10.0\ mL + 2 \times 0.100\ mmol/mL \times 10.0\ mL = 3.00\ mmol$

$\underline{M}_{NO_3^-} = 3.00\ mmol/30.0\ mL = 0.100\ mmol/mL$

$mmol\ K^+ = mmol\ KNO_3 + 2 \times mmol\ K_2SO_4$

$= 0.100\ mmol/mL \times 10.0\ mL + 2 \times 0.100\ mmol/mL \times 10.0\ mL = 3.00\ mmol$

$\underline{M}_{K^+} = 3.00\ mmol/30.0\ mL = 0.100\ mmol/mL$

$mmol\ SO_4^{2-} = mmol\ K_2SO_4 = 0.100\ mmol/mL \times 10.0\ mL = 1.00\ mmol$

$\underline{M}_{SO_4^{2-}} = 1.00\ mmol/30.0\ mL = 0.0333\ mmol/mL$

17. $10.0\ mmol/L = 0.0100\ mmol/mL$

$0.0100\ mmol/mL \times 0.147\ g/mmol = 0.00147\ g\ CaCl_2 \cdot 2H_2O/mL$

18. (a) $(10.0\ g/250\ mL)/(98.1\ g/mol) \times 1000\ mL/L = 0.408\ \underline{M}\ H_2SO_4$

(b) $(6.00\ g/500\ mL)/(40.0\ g/mol) \times 1000\ mL/L = 0.300\ \underline{M}\ NaOH$

(c) $(25.0\ g/L)/(170\ g/mol) = 0.147\ \underline{M}\ AgNO_3$

19. (a) $(0.100\ mol/L)(142\ g/mol)(0.500\ L) = 7.10\ g\ Na_2SO_4$

(b) $(0.250\ mol/L)(392\ g/mol)(0.500\ L) = 49.0\ g\ Fe(NH_4)_2(SO_4)_2 \cdot 6H_2O$

(c) $(0.667\ mol/L)(328\ g/mol)(0.500\ L) = 109\ g\ Ca(C_9H_6ON)_2$

20. (a) $(250\ mL)(0.100\ mmol/mL)(56.1\ mg/mmol)(1.00 \times 10^{-3}\ g/mg) = 1.40\ g\ KOH$

(b) $(1.00\ L)(0.0275\ mol/L)(294\ g/mol) = 8.08\ g\ K_2Cr_2O_7$

(c) $(500\ mL)(0.0500\ mmol/mL)(160\ mg/mmol)(1.00 \times 10^{-3}\ g/mg) = 4.00\ g\ CuSO_4$

21. $(0.380\ g/g \times 1.19\ g/mL)/36.5\ g/mol \times 1000\ mL/L = 12.4\ mol/L$ in stock sol'n

$12.4\ mmol/mL \times X\ mL = 0.100\ mmol/mL \times 1000\ mL$

$X = 8.06\ mL$ must be diluted

22. (a) $(0.700\ g/g \times 1.668\ g/mL)/100.5\ g/mol \times 1000\ mL/L = 11.6\ mol/L$

(b) $(0.690\ g/g \times 1.409\ g/mL)/63.01\ g/mol \times 1000\ mL/L = 15.4\ mol/L$

(c) $(0.850\ g/g \times 1.689\ g/mL)/98.0\ g/mol \times 1000\ mL/L = 14.6\ mol/L$

(d) $(0.995\ g/g \times 1.051\ g/mL)/60.05\ g/mol \times 1000\ mL/L = 17.4\ mol/L$

(e) $(0.280\ g/g \times 0.898\ g/mL)/17.03\ g/mol \times 1000\ mL/L = 14.8\ mol/L$

23. $6.0 \times 10^{-6}\ mol/250\ mL = 24 \times 10^{-6}\ mol/L\ Na_2SO_4$

$= 48 \times 10^{-6}\ mol/L$ of Na^+ and $24 \times 10^{-6}\ mol/L$ of SO_4^{2-}

$48 \times 10^{-6}\ mol/L \times 23 \times 10^3\ mg/mol = 1.1\ mg/L\ Na^+$

$24 \times 10^{-6}\ mol/L \times 96 \times 10^3\ mg/mol = 2.3\ mg/L\ SO_4^{2-}$

24. $325\ mg/L \times 0.100\ L = 32.5\ mg\ K^+$

$(32.5 \times 10^{-3}\ g)/(39.1\ g/mol) = 8.31 \times 10^{-4}\ mol\ K^+ = 8.31 \times 10^{-4}\ mol\ (C_6H_5)_4B^-$

$8.31 \times 10^{-4}\ mol \times 319\ g/mol \times (1000\ mL/L)/(250\ mL) \times 10^3\ mg/g$

$= 1.06 \times 10^3\ mg/L\ (C_6H_5)_4B^-$

25. $g/mL = ppm \times 10^{-6} = 1.00 \times 10^{-6}\ g/mL = 1.00 \times 10^{-3}\ g/L$

(a) $(1.00 \times 10^{-3}\ g/L)/(170\ g/mol) = 5.88 \times 10^{-6}\ mol/L\ AgNO_3$

(b) $(1.00 \times 10^{-3}$ g/L)/(342 g/mol) = 2.92×10^{-6} mol/L $Al_2(SO_4)_3$

(c) $(1.00 \times 10^{-3}$ g/L)/(44.0 g/mol) = 2.27×10^{-5} mol/L CO_2

(d) $(1.00 \times 10^{-3}$ g/L)/(633 g/mol) = 1.58×10^{-6} mol/L $(NH_4)_4Ce(So_4)_4.2H_2O$

(e) $(1.00 \times 10^{-3}$ g/L)/(36.5 g/mol) = 2.73×10^{-5} mol/L HCl

(f) $(1.00 \times 10^{-3}$ g/L)/(100 g/mol) = 1.00×10^{-5} mol/L $HClO_4$

26. 2.50×10^{-4} mol/L x 10^3 mmol/mol = 0.250 mmol/L

(a) 0.250 mmol/L x 40.1 mg/mmol = 10.0 mg/L = ppm Ca^{2+}

(b) 0.250 mmol/L x 111 mg/mmol = 27.8 mg/L $CaCl_2$

(c) 0.250 mmol/L x 63.0 mg/mmol = 15.8 mg/L HNO_3

(d) 0.250 mmol/L x 65.1 mg/mmol = 16.3 mg/L KCN

(e) 0.250 mmol/L x 54.9 mg/mmol = 13.7 mg/L Mn^{2+}

(f) 0.250 mmol/L x 119 mg/mmol = 29.8 mg/L MnO_4^-

27. 1.00 ppm Fe^{2+} = 1.00 mg/L = 0.00100 g/L

(0.00100 g/L)/(55.8 g/mol) = 1.79×10^{-5} mol/L

1 mol $FeSO_4.(NH_4)_2SO_4.6H_2O$ contains 1 mol of Fe.

.·. require 1.79×10^{-5} mol of this

1.79×10^{-5} mol x 392 g/mol = 0.00702 g $FeSO_4.(NH_4)_2SO_4.6H_2O$

It is simpler to multiply the weight by the ratio of the formula weights:

g $FeSO_4.(NH_4)_2SO_4.6H_2O$ = g Fe x [f.w. $FeSO_4.(NH_4)_2SO_4.6H_2O$/at. wt. Fe]

= 0.001 x (392)/(55.8) = 0.00702 g

The number of iron atoms in the numerator and denominator of the formula weight ratio must be the same.

28. (a) % Cr_2O_3 = [0.560 mg/456 mg] x 100% = 0.123%

(b) ppt Cr_2O_3 = [0.560 mg/456 mg] x 1000°/₀₀ = 1.23°/₀₀

(c) ppm Cr_2O_3 = [0.560 mg/456 mg] x 10^6 = 1.23 x 10^3 ppm

29. 100 ppm x 10^{-6} = 1.00 x 10^{-4} g/mL = 0.100 g/L

(a) (0.100 g/L)/(23.0 g/mol) = 4.35 x 10^{-3} mol/L Na^+

The molar contraction of NaCl is the same as that of Na^+ (and Cl^-).

4.35 x 10^{-3} mol/L x 58.4 g/mol = 0.254 g/L NaCl

(b) (0.100 g/L)/(35.5 g/mol) = 2.82 x 10^{-3} mol/L Cl^- = NaCl

2.82 x 10^{-3} mol/L x 58.4 g/mol = 0.165 g/L NaCl

30. 250 ppm x 10^{-6} = 2.50 x 10^{-4} g/mL = 0.250 g/L K^+

(0.250 g/L)/(39.1 g/mol) = 6.39 x 10^{-3} mol/L K^+ = KCl = Cl^-

The millimoles before and after dilution are equal. So

6.39 x 10^{-3} mmol/mL x X mL = 1.00 x 10^{-3} mmol/mL x 1000 mL

X = 156 mL KCl required

31. 500 ppm x 10^{-6} = 5.00 x 10^{-4} g/mL = 0.500 g/L $KClO_3$. There is 1 $K^+/KClO_3$.

0.500 g $KClO_3$ x $(K/KClO_3)$ = 0.500 g x (39.1)/(122) = 0.160 g K^+

32. 12.5 mL x M = 500 mL x 0.1225 M

M = 5.00 M

33. Let x = mL 0.50 M H_2SO_4

0.35 M x (65 + x) mL = 0.20 M x 65 mL + 0.50 M x x mL

x = 65 mL

34. mmol NaOH = 50 x 0.10 = 5.0 mol

mL H_2SO_4 to neutralize NaOH = 5.0/(0.10 x 2) = 25 mL

Let x = additional mL H_2SO_4 required

0.050 M x (50 + 25 + x) mL = 0.10 M x x mL

x = 75 mL

∴ total volume = 25 + 75 = 100 mL

35. The most concentrated solution, 1.00×10^{-4} $\underline{M}$, would require 1:1,000 dilution of the stock solution, and you can not do this with the glassware provided. Prepare a diluted stock solution. The most dilute possible is 1:100 (1 mL diluted to 100 mL):

$$0.100\ \underline{M} \times 1.00\ \text{mL} = \underline{M}_A \times 100\ \text{mL}$$

$$\underline{M}_A = 1.00 \times 10^{-3}\ \underline{M}$$

This can be diluted appropriately to give the desired concentrations.

1.00×10^{-5} $\underline{M}$: dilute 1 mL to 100 mL (1:100)

2.00×10^{-5} $\underline{M}$: dilute 2 mL to 100 mL (1:50)

5.00×10^{-5} $\underline{M}$: dilute 5 mL to 100 mL (1:20)

1.00×10^{-4} $\underline{M}$: dilute 10 mL to 100 mL (1:10)

36. The dilution factor for the original solution is 1:5 (50 mL aliquot diluted to 250 mL). Hence, the concentration in the original solution is $5 \times (1.25 \times 10^{-5}\ \underline{M}) = 6.25 \times 10^{-5}\ \underline{M}$. We have 250 mL at this concentration.

$$\text{mmol}_{Mn} = 6.25 \times 10^{-5}\ \underline{M} \times 250\ \text{mL} = 0.0156\ \text{mmol}$$

$$\text{mg}_{Mn} = 0.0156\ \text{mmol} \times 54.9\ \text{mg/mmol} = 0.858\ \text{mg} = 8.58 \times 10^{-4}\ \text{g}$$

$$\%\ Mn = [(8.58 \times 10^{-4}\ \text{g})/(0.500\ \text{g})] \times 100\% = 0.172\%$$

37. The neutralization reaction is $Na_2CO_3 + H_2SO_4 \longrightarrow Na_2SO_4 + 2H_2O$.

$$\therefore\ \%\ Na_2CO_3 = \frac{\underline{M}_{H_2SO_4} \times mL_{H_2SO_4} \times 1(\text{mmol } H_2SO_4/\text{mmol } Na_2CO_3) \times f.w._{Na_2CO_3}}{\text{mg sample}} \times 100\%$$

$$98.6\% = \frac{\underline{M}_{H_2SO_4} \times 36.8 \times 1 \times 106.0}{678} \times 100\%$$

$$\underline{M}_{H_2SO_4} = 0.171\ \underline{M}$$

38. $40\ \text{mL} \times 0.10\ \underline{M} = 4.0\ \text{mmol NaOH} = \text{mmol sulfamic acid}$

4.0 mmol = (x mg)/(97 mg/mmol); x = 390 mg sulfamic acid

39. % H_3A = {[0.1087 M x 38.31 mL x 1/3 (mmol H_3A/mmol NaOH)

x 192.1 mg/mmol]/(267.8 mg)} x 100% = 99.57%

40. mmol Ca^{2+} = mmol EDTA

mg Ca^{2+} = 1.87 x 10^{-4} M x 2.47 mL x 40.1 mg/mmol = 0.0185 mg

(200 μL)/(1000 μL/mL) = 0.200 mL

(0.0185 mg/0.200 mL) x 100 mL/dL = 9.25 mg/dL

41. (a) % Cl = {[0.100 M x 27.2 mL x 1 (mmol Cl^-/mmol Ag^+)

x 35.5 mg/mmol]/(372 mg)} x 100% = 26.0%

(b) % $BaCl_2.2H_2O$ = {[0.100 M x 27.2 mL x 1/2 (mmol $BaCl_2$/mmol Ag^+)

x 244 mg/mmol]/372 mg} x 100% = 89.2%

42. Each $Cr_2O_7^{2-}$ reacts with $6Fe^{2+}$ (≡ $3Fe_2O_3$), .˙. % Fe_2O_3 =

[($M_{Cr_2O_7^{2-}}$ x $mL_{Cr_2O_7^{2-}}$ x 3(mmol Fe_2O_3/mmol $Cr_2O_7^{2-}$) x f.w.$_{Fe_2O_3}$

(mg/mmol)/(mg sample)] x 100% = [(0.0150 x 35.6 x 3 x 160)/(1680)] x 100%

= 15.3%

43. 1 Ca = 1 $H_2C_2O_4$ = 2/5 MnO_4^-

.˙. 1 MnO_4^- = 5/2 CaO

% CaO = [(35.6 mL x 0.0200 mmol/mL x 5/2 x 56.1 mg/mmol)/(2000 mg)] x 100%

= 4.99%

44. $M_{MnO_4^-}$ = (4680 mg)/($KMnO_4$ x 500 mL) = (4680 mg)/(158.0 mg/mmol) x 500 mL)

= 0.0592 mmol/mL

1 MnO_4^- = 2.5 Fe_2O_3

.˙. 35.6% = [(0.0592 mmol/mL x mL x 2.5 x Fe_2O_3 mg/mmol)/(500 mg)] x 100%

$= [(0.0592 \times mL \times 2.5 \times 159.7)/(500\ mg)] \times 100\%$

$mL = 7.53$

45. Let $a = \%\ BaCl_2 = mL\ AgNO_3$

$1\ Ag = 1/2\ BaCl_2$

$\therefore\ a = [(0.100\ \underline{M} \times a\ mL \times 1/2 \times BaCl_2)/(mg)] \times 100\%$

$1 = [(0.100 \times 1/2 \times 208)/(mg)] \times 100\%$

$mg = 1.04 \times 10^3$ mg sample

46. $\%\ AlCl_3 = \{[0.100\ \underline{M} \times 48.6\ mL \times 1/3\ (mmol\ AlCl_3/mmol\ Ag^+)$

$\times 133\ mg/mmol]/(250\ mg)\} \times 100\% = 86.2\%$

mmol Al in 350 mg $= [(350\ mg \times 0.862)/133\ (mg\ AlCl_3/mmol)]$

$= 2.27$ mmol Al (= mmol $AlCl_3$)

$2.27\ mmol = 0.100 \times x\ mL$

$x = 22.7$ mL EDTA

47. $100\% = [(0.1027\ \underline{M} \times 28.78\ mL \times f.w._{HA})/(425.2\ mg)] \times 100\%$

$f.w._{HA} = 143.9$

48. (462 mg AgCl)/143 mg/mmol) = 3.23 mmol AgCl = mmol HCl

$\underline{M}_{HCl} \times 25.0\ mL = 3.23\ mmol$

$\underline{M}_{HCl} = 0.129\ \underline{M}$

$\%\ Zn(OH)_2 = \{[0.129\ \underline{M} \times 37.8\ mL \times 1/2\ (mmol\ Zn(OH)_2/mmol\ HCl)$

$\times 99.4\ mg/mmol]/(287\ mg)\} \times 100\% = 84.4\%$

49. mmol $KHC_2O_4.H_2C_2O_4.2H_2O$ = 1/3 mmol NaOH = 1/3 x 46.2 mL x 0.100 mmol/mL

= 1.54 mmol

mmol $C_2O_4^{2-}$ = 2 x mmol $KHC_2O_4.H_2C_2O_4.2H_2O$ = 2 x 1.54 = 3.08 mmol

Each $C_2O_4^{2-}$ = 2/5 MnO_4^-

∴ mmol $KMnO_4$ = 2/5 x 3.08 = 1.23 mmol

0.100 mol/mL x $mL_{MnO_4^-}$ = 1.23 mmol

$mL_{MnO_4^-}$ = 12.3 mL

50. mmol Na_2CO_3 = 1/2 mmol HCl reacted

∴ % Na_2CO_3 = {[(0.100 x 50.0 - 0.100 x 5.6) x 1/2 x 106.0]/(500 mg)} x 100%

= 47.1%

51. 1 MnO_4^- = 5/2 H_2O_2; 1 Fe^{2+} = 1/5 MnO_4^-

∴ mmol H_2O_2 = mmol MnO_4^- reacted x 5/2

= (mmol MnO_4^- taken - mmol MnO_4^- unreacted) x 5/2

= (mmol MnO_4^- taken - mmol Fe^{2+} x 1/5) x 5/2

∴ % H_2O_2 =

{[($\underline{M}_{MnO_4^-}$ x $mL_{MnO_4^-}$ - $\underline{M}_{Fe^{2+}}$ x $mL_{Fe^{2+}}$ x 1/5) x 5/2 x $f.w._{H_2O_2}$]/(mg_{sample})} x 100% = {[(25.0 x 0.0215 - 5.10 x 0.112 x 1/5) x 5/2 x 34.0]/(587)} x 100%

= 6.13%

52. mmol excess I_2 = 1/2 mmol $S_2O_3^{2-}$

mmol H_2S = mmol reacted I_2

∴ mg S = (0.00500 $\underline{M}$ x 10.0 mL - 1/2 x 0.00200 $\underline{M}$ x 2.6 mL) x 32.06 = 1.52 mg

53. The reaction is Ba^{2+} + H_2EDTA^{2-} = $Ba\text{-}EDTA^{2-}$ + $2H^+$.

∴ mol EDTA = mmol BaO

0.100 mmol/mL x 1 mL = (mg BaO)/(BaO) = (mg BaO)/153 mg/mmol)

Titer = 15.3 mg BaO per milliliter EDTA

54. 1 Fe reacts with 1/5 MnO_4^-

$\therefore$ 5/2 mmol MnO_4^- = mmol Fe_2O_3

5/2 x 0.0500 mmol/mL x 1 mL = (mg Fe_2O_3)/(Fe_2O_3) = (mg Fe_2O_3)/(159.7 mg/mmol)

Titer = 20.0 mg Fe_2O_3 per milliliter of $KMnO_4$

55. mmol Ag^+ = mmol X^-

$\underline{M}$ x 1 mL = (22.7 mg)/(Cl) = (22.7 mg)/(35.4 mg/mmol)

$\underline{M}_{AgNO_3}$ = 0.641 $\underline{M}$

0.641 $\underline{M}$ x 1 mL = (mg Br)/(Br) = (mg Br)/(79.9 mg/mmol)

Titer = 51.2 mg Br/mL

or Titer = 22.7 mg x (Br/Cl) = 22.7 mg x (79.9/35.4) = 51.2 mg Br/mL

56. (a) (36.46 g/mol)/(1 eq/mol) = 36.46 g/eq

(b) (171.34 g/mol)/(2 eq/mol) = 85.67 g/eq

(c) (389.91 g/mol)/(1 eq/mol) = 389.91 g/eq

(d) (82.08 g/mol)/(2 eq/mol) = 41.04 g/eq

(e) (60.05 g/mol)/(1 eq/mol) = 60.05 g/eq

57. (a) (0.250 eq/L)/(1 mol/eq) = 0.250 mol/L

(b) (0.250 eq/L)/(1/2 mol/eq) = 0.125 mol/L

(c) (0.250 eq/L)/(1 mol/eq) = 0.250 mol/L

(d) (0.250 eq/L)/(1/2 mol/eq) = 0.125 mol/L

(e) (0.250 eq/L)/(1 mol/eq) = 0.250 mol/L

58. (a) eq wt = (128.1 g/mol)/(1 eq/mol) = 128.1 g/eq

(b) Each C is oxidized from +3 to +4, so the change is 2 electrons/$HC_2O_4^-$.

eq wt = (128.1 g/mol)/(2 eq/mol) = 64.05 g/eq

59. eq wt = (216.6 g/mol)/(2 eq/mol) = 108.3 g/eq

60. (a) 151.91/1 = 151.91 g/eq (b) 34.08/2 = 17.04 g/eq

(c) 34.01/2 = 17.00 g/eq (d) 34.01/2 = 17.00 g/eq

61. meq = (mg/[eq wt (mg/meq)]) = (500.0/[244.3/2(mg/meq)]) = 4.093 meq

62. 7.82 g NaOH/(40.0 g/eq) = 0.196 eq

9.26 g $Ba(OH)_2$/(171/2 g/eq) = 0.108 eq

$\underline{N}$ = [(0.196 + 0.108) eq/(500 mL)] x 1000 mL/L = 0.608 eq/L

63. Each As undergoes 2 electron change. Therefore,

eq wt = As_2O_3/4 = 197.8/4 = 49.45 g/eq

0.1000 $\underline{N}$ = 0.1000 eq/L; 0.1000 eq/L x 49.45 g/eq = 4.945 g/L

64. 2.73 g x 0.980 = 2.68 g $KHC_2O_4.H_2C_2O_4$

2.68 g/(218/3 g/eq) = 0.0369 eq

1.68 $KHC_8H_4O_4$/(204 g/eq) = 0.00823 eq

$\underline{N}$ = [(0.0369 + 0.00823) eq/(250 mL)] x 1000 mL/L = 0.180 eq/L

65. The carbon in each oxalate ($C_2O_4^{2-}$) is oxidized from +3 to +4, releasing 2 electrons/$C_2O_4^{2-}$. Therefore, the equivalent weight of $KHC_2O_4.H_2C_2O_4.2H_2O$ as a reducing agent is one-fourth its formula weight. ∴ $\underline{N}_{red} = \underline{N}_{acid} \times 4/3$ = 0.200 x 4/3 = 0.267 $\underline{N}$.

66. Assume 1.00 $\underline{N}$ as acid, therefore, 3.62 $\underline{N}$ as reducing agent. Acidity is due to $KHC_2O_4.H_2C_2O_4$, and so the concentration of this is 1.00 $\underline{N}$ as an acid or 1.00(4/3) = 1.33 $\underline{N}$ as a reducing agent. If we have 1 L of solution, the normality and equivalents are equal.

∴ 1.33 eq $KHC_2O_4.H_2C_2O_4$ + x eq $Na_2C_2O_4$ = 3.62 eq (as reducing agent)

x = 2.29 eq $Na_2C_2O_4$

$1.33\ eq \times (218\ g/4)\ g/eq = 72.5\ g\ KHC_2O_4 \cdot H_2C_2O_4$

$2.29\ eq \times (134/2)\ g/eq = 153\ g\ Na_2C_2O_4$

$\therefore$ ratio $= 72.5/153 = 0.474\ g\ KHC_2O_4 \cdot H_2C_2O_4\ /g.\ Na_2C_2O_4$

67. meq = $\underline{N}$ x mL = 0.100 meq/mL x 1000 mL = 100.0 meq

mg = meq x eq wt (mg/meq) = 100.0 meq x (294.2/6)(mg/meq)

= 4903 mg (4.903 g)

68. (300 mg/dL)/(0.1 L/dL) = 3.00×10^3 mg/L

(3.00×10^3 mg/L)/(35.5 mg/meq) = 84.5 meq/L

69. [(5.00 meq/L) x (40.1/2 mg/meq)]/(10 dL/L) = 10.0 mg/dL

70. There are 150 mmol/L of NaCl. $\therefore$ 150 mol/L x 0.0584 g/mmol = 8.76 g/L.

71. g Mn = g Mn_3O_4 x ($3Mn/Mn_3O_4$) g Mn/g Mn_3O_4

= 2.58 g Mn_3O_4 x [3(54.9)/228.8] = 1.85_7 g Mn

72. (a) g Zn = g $Zn_2Fe(CN)_6$ x ($2Zn/Zn_2Fe(CN)_6$) g Zn/g $Zn_2Fe(CN)_6$

= 0.348 x [2(65.4)/342.7] = 0.348 x 0.369 (g Zn/g $Zn_2Fe(CN)_6$) = 0.132_8 g Zn

(0.369 is the gravimetric factor)

(b) g $Zn_2Fe(CN)_6$ = g Zn x (1/2 $Zn_2Fe(CN)_6$/Zn) g $Zn_2Fe(CN)_6$/g Zn

= 0.500 g Zn x [1/2(342.7/65.4) = 1.31_0 g $Zn_2Fe(CN)_6$

73. 3 Mn/Mn_3O_4 = 3(54.938)/228.81 = 0.7203_1 g Mn/g Mn_3O_4

$3Mn_2O_3/2Mn_3O_4$ = 3(157.88)/2(228.81) = 2.0700 g Mn_2O_3/g Mn_3O_4

$Ag_2S/BaSO_4$ = 247.80/233.40 = 1.0617_0 g Ag_2S/g $BaSO_4$

$CuCl_2/2AgCl$ = 134.45/2(143.32) = 0.46906 g $CuCl_2$/g AgCl

MgI_2/PbI_2 = 278.12/261.00 = 1.0656 g MgI_2/g PbI_2

CHAPTER 6

1. $K = [C][D]/[A][B] = 2.0 \times 10^{3}$

   At equilibrium,

   $[A] = x$

   $[B] = (0.80 - 0.30) + x = 0.50 + x \approx 0.50$ $\underline{M}$

   $[C] = [D] = 0.30 - x \approx 0.30$ $\underline{M}$

   $[(0.30)(0.30)]/[(x)(0.50)] = 2.0 \times 10^{3}$

   $x = 9.0 \times 10^{-5}$ $\underline{M} = [A]$ (We were justified in neglecting x above.)

2. $A + B = 2C$

   $[C]^{2}/[A][B] = 5.0 \times 10^{6}$

   The reaction is limited by the amount of A (0.40 $\underline{M}$) available to react. At equilibrium:

   $[A] = x$

   $[B] = (0.70 - 0.40) + x = 0.30 + x \approx 0.30$

   $[C] = 2(0.40) - x = 0.80 - x \approx 0.80$

   $(0.80)^{2}/(x)(0.30) = 5.0 \times 10^{6}$

   $x = 4.3 \times 10^{-7}$ $\underline{M} = [A]$

3. $HA = H^{+} + A^{-}$

   $[H^{+}][A]/[HA] = K_{eq} = 1.0 \times 10^{-3}$

   At equilibrium,

   $[H^{+}] = [A^{-}] = x$

   $[HA] = 1.0 \times 10^{-3} - x$. Since $[HA] < 100 \times K_{eq}$, we can't neglect x. Solve for the quadratic formula.

   $(x)(x)/(1.0 \times 10^{-3} - x) = 1.0 \times 10^{-3}$

$x^2 + 1.0 \times 10^{-3}x - 1.0 \times 10^{-6} = 0$

$x = (-1.0 \times 10^{-3}) \pm \sqrt{(1.0 \times 10^{-3})^2 + 4.0 \times 10^{-6}}/2 = 6._0 \times 10^{-4}$ M

% dissociated = $[(6.0 \times 10^{-4})/(1.0 \times 10^{-3})] \times 100\% = 60\%$

4. $HCN = H^+ + CN^-$

$1.0 \times 10^{-3} - x \quad x \quad x$

Neglect x compared to 1.0×10^{-3} ($C >> 200 \times K_{eq}$)

$[H^+][CN^-]/[HCN] = 7.2 \times 10^{-10}$

$(x)(x)/(1.0 \times 10^{-3}) = 7.2 \times 10^{-10}$

$x = 8.5 \times 10^{-7}$ M

% dissociated = $[(8.5 \times 10^{-7})/(1.0 \times 10^{-3})] \times 100\% = 0.085\%$

5. $HA = H^+ + A^-$

At equilibrium,

$[H^+] = x$

$[A^-] = 1.0 \times 10^{-2} + x \approx 1.0 \times 10^{-2}$

$[HA] = 1.0 \times 10^{-3} - x \approx 1.0 \times 10^{-3}$ (A^- will suppress the dissociation, so let's assume x is now small.)

$[H^+][A^-]/[HA] = 1.0 \times 10^{-3}$

$(x)(1.0 \times 10^{-2})/(1.0 \times 10^{-3}) = 1.0 \times 10^{-3}$

$x = 1.0 \times 10^{-4}$ M = concentration dissociated

% dissociated = $[(1.0 \times 10^{-4})/(1.0 \times 10^{-3})] \times 100\% = 10\%$

6. $H_2S = H^+ + HS^-$ $\quad K_1 = [H^+][HS^-]/[H_2S] = 9.1 \times 10^{-8}$

$HS^- = H^+ + S^{2-}$ $\quad K_2 = [H^+][S^{2-}]/[HS^-] = 1.2 \times 10^{-15}$

Overall:

$H_2S = 2H^+ + S^{2-}$ $\quad K = [H^+]^2[S^{2-}]/[H_2S] = K_1K_2 = (9.1 \times 10^{-8})(1.2 \times 10^{-15})$

$= 1.1 \times 10^{-22}$

7. The initial analytical concentrations after mixing but before reaction, are

$[Fe^{2+}] = 0.06\ \underline{M}$

$[Cr_2O_7^{2-}] = 0.01\ \underline{M}$

$[H^+] = 1.14\ \underline{M}$

There are stoichiometrically equal concentrations of Fe^{2+} and $Cr_2O_7^{2-}$. At equilibrium, 0.02 mol/L $Cr_2O_7^{2-}$ has reacted with 0.14 mol/L H^+, leaving 1.00 mol/L H^+ (plus the amount from the reverse equilibrium reaction), so

$6Fe^{2+}$ +	$Cr_2O_7^{2-}$ +	$14H^+$	$= 6Fe^{3+}$ +	$2Cr^{3+}$ +	$7H_2O$
6x	x	1.00 + 14x	0.06 − 6x	0.02 − 2x	
		≈ 1.00	≈ 0.06	≈ 0.02	

$([Fe^{3+}]^6[Cr^{3+}]^2)/([Fe^{2+}]^6[Cr_2O_7^{2-}][H^+]^{14}) = 1 \times 10^{57}$

$[(0.06)^6\ (0.02)^2]/[(6x)^6\ (x)\ (1.00)^{14}] = 1 \times 10^{57}$

$x^7 = 4 \times 10^{-73} = 40{,}000 \times 10^{-77}$

$x = [Cr_2O_7^{2-}] = 5 \times 10^{-11}\ \underline{M}$

(See Appendix B for a review of calculating odd roots)

$[Fe^{2+}] = 6x = 3 \times 10^{-10}\ \underline{M}$

8. (a) Equilibria:

$\underline{Bi_2S_3} \rightleftharpoons 2Bi^{3+} + 3S^{2-}$

$S^{2-} + H^+ \rightleftharpoons HS^-$

$HS^- + H^+ \rightleftharpoons H_2S$

$H_2O \rightleftharpoons H^+ + OH^-$

$3[Bi^{3+}] + [H^+] = 2[S^{2-}] + [HS^-] + [OH^-]$

(b) *Equilibria:*

$$Na_2S \rightarrow 2Na^+ + S^{2-}$$

$$S^{2-} + H^+ \rightleftharpoons HS^-$$

$$HS^- + H^+ \rightleftharpoons H_2S$$

$$H_2O \rightleftharpoons H^+ + OH^-$$

$$[Na^+] + [H^+] = 2[S^{2-}] + [HS^-] + [OH^-]$$

9. Mass balance

$$(Cd): \; 0.100 = [Cd^{2+}] + [Cd(NH_3)^{2+}] + [Cd(NH_3)_2^{2+}] + [Cd(NH_3)_3^{2+}] + [Cd(NH_3)_4^{2+}]$$

$$(N): \; 0.400 = [NH_4^+] + [NH_3] + [Cd(NH_3)^{2+}] + 2[Cd(NH_3)_2^{2+}] + 3[Cd(NH_3)_3^{2+}] + 4[Cd(NH_3)_4^{2+}]$$

$$(Cl): \; 0.200 = [Cl^-]$$

Charge balance:

$$2[Cd^{2+}] + [H^+] + 2[Cd(NH_3)^{2+}] + 2[Cd(NH_3)_2^{2+}] + 2[Cd(NH_3)_3^{2+}] + 2[Cd(NH_3)_4^{2+}] =$$
$$[Cl^-] + [OH^-]$$

10. (a) Charge balance (C.B.): $[H^+] = [NO_2^-] + [OH^-]$, $\therefore [NO_2^-] = [H^+] - [OH^-]$

(b) Mass balance (M.B.): $0.2 = [CH_3COO^-] + [CH_3COOH]$ (1)

C.B.: $[H^+] = [CH_3COO^-] + [OH^-]$ (2)

Combining Equations (1) and (2) : $[CH_3COOH] = 0.2 - [CH_3COO^-] = 0.2 - ([H^+] - [OH^-]) =$
$0.2 - [H^+] + [OH^-]$

(c) M.B. : $0.1 = [H_2C_2O_4] + [HC_2O_4^-] + [C_2O_4^{2-}]$ (3)

C.B.: $[H^+] = [OH^-] + [HC_2O_4^-] + 2[C_2O_4^{2-}]$ (4)

Combining Equations (3) and (4) : $[H_2C_2O_4] = 0.1 - ([HC_2O_4^-] + [C_2O_4^{2-}]) =$

$0.1 - ([H^+] - [OH^-] - [C_2O_4^{2-}]) = 0.1 - [H^+] + [OH^-] + [C_2O_4^{2-}]$

(d) M.B. : $0.1 = [CN^-] + [HCN] = [K^+]$ (5)

C.B.: $[K^+] + [H^+] = [OH^-] + [CN^-]$ (6)

Combining Equations (5) and (6) : $[HCN] = 0.1 - [CN^-] = 0.1 - (0.1 + [H^+] - [OH^-]) =$

$[OH^-] - [H^+]$

(e) M.B. : $0.1 = [H_3PO_4] + [H_2PO_4^-] + [HPO_4^{2-}] + [PO_4^{3-}]$ (7)

C.B.: $[Na^+] + [H^+] = 0.3 + [H^+] = [OH^-] + [H_2PO_4^-] + 2[HPO_4^{2-}] + 3[PO_4^{3-}]$ (8)

or $0.3 = [OH^-] - [H^+] + [H_2PO_4^-] + 2[HPO_4^{2-}] + 3[PO_4^{3-}]$ (8a)

Multiplying Equation (7) by 3 and combining the result with Equation (8a),

$$3[H_3PO_4] + 3[H_2PO_4^-] + 3[HPO_4^{2-}] + 3[PO_4^{3-}] = [OH^-] - [H^+] + [H_2PO_4^-] + 2[HPO_4^{2-}] + 3[PO_4^{3-}]$$

or $$[H_2PO_4^-] = \frac{[OH^-] - [H^+] - [HPO_4^{2-}] - 3[H_3PO_4]}{2}$$

(f) M.B.: $0.1 = [HSO_4^-] + [SO_4^{2-}]$ (because $[H_2SO_4] = 0$) (9)

C.B. $[H^+] = [OH^-] + [HSO_4^-] + 2[SO_4^{2-}]$ (10)

Combining Equations (9) and (10), we have

$$[HSO_4^-] = 0.1 - [SO_4^{2-}] = 0.1 - ([H^+] - [OH^-] - [HSO_4^-])/2$$

or $$[HSO_4^-] = 0.2 - [H^+] + [OH^-]$$

11. If S is the molar solubility of BaF_2, then

$S = [Ba^{2+}]$

$2S = [F^-] + [HF] + 2[HF_2^-]$ $\quad\therefore\quad$ $[F^-] + [HF] + 2[HF_2^-] = 2[Ba^{2+}]$

12. $2[Ba^{2+}] = 3([PO_4^{3-}] + [HPO_4^{2-}] + [H_2PO_4^-] + [H_3PO_4])$

13. *Equilibria:*

$HOAc \rightleftharpoons H^+ + OAc^-$

$H_2O \rightleftharpoons H^+ + OH^-$

Equilibrium expressions

$K_a = [H^+][OAc^-]/[HOAc] = 1.75 \times 10^{-5}$ (1)

$K_w = [H^+][OH^-] = 1.00 \times 10^{-14}$ (2)

Mass balance expressions

$C_{HOAc} = [HOAc] + [OAc^-] = 0.100$ M (3)

$[H^+] = [OAc^-] + [OH^-]$ (4)

Charge balance expression

$[H^+] = [OAc^-] + [OH^-]$ (5)

Number of expressions vs. number of unknowns

There are four unknowns ([HOAc], [OAc$^-$], [H$^+$], [OH$^-$]) and four expressions (two equilibrium and two mass balance — the charge balance expression is the same as (4)).

Simplifying assumptions:

In acid solution [OH$^-$] << [H$^+$]. Assume [OAc$^-$] << [HOAc], since K_a < 0.01 C.

From (4): $[OAc^-] \approx [H^+]$

From (3): $[HOAc] \approx 0.100\ \underline{M}$

From (1):

$[H^+]^2/0.100 = 1.75 \times 10^{-5}$

$[H^+] = 3.77 \times 10^{-3}$

pH = 2.43

14. (a) $\mu = ([Na^+](1)^2 + [Cl^-](1)^2)/2 = [(0.30)(1) + (0.30)(1)]/2 = 0.30$

(b) $\mu = ([Na^+](1)^2 + [SO_4^{2-}](2)^2)/2 = [(0.60)(1) + (0.30)(4)]/2 = 0.90$

(c) $\mu = ([Na^+](1)^2 + [Cl^-](1)^2 + [K^+](1)^2 + [SO_4^{2-}](2)^2)/2$

$= [(0.30)(1) + (0.30)(1) + (0.40)(1) + (0.20)(4)]/2 = 0.90$

(d) $\mu = ([Al^{3+}](3)^2 + [SO_4^{2-}](2)^2 + [Na^+](1)^2)/2$

$= [(0.40)(9) + (0.60 + 0.10)(4) + (0.20)(1)]/2 = 3.3$

15. (a) $\mu = ([Zn^{2+}](2)^2 + [SO_4^{2-}](2)^2)/2 = [10.20(4) + (0.20)(4)]/2 = 0.80$

(b) $\mu = ([Mg^{2+}](2)^2 + [Cl^-](1)^2)/2 = [(0.40)(4) + (0.80)(1)]/2 = 1.20$

(c) $\mu = ([La^{3+}](3)^2 + [Cl^-](1)^2)/2 = [(0.50)(9) + (1.50)(1)]/2 = 3.0$

(d) $\mu = ([K^+](1)^2 + [Cr_2O_7^{2-}](2)^2)/2 = [(2.0)(1) + (1.0)(4)]/2 = 3.0$

(e) $\mu = ([Tl^{3+}](3)^2 + [Pb^{2+}](2)^2 + [NO_3^-](1)^2)/2$

$= [(1.0)(9) + (1.0)(4) + (5.0)(1)]/2 = 9.0$

16. μ = [NaCl] = 1.00×10^{-3}. This is less than 0.01, so use Equation 6.20.

$$-\log f_{Na^+} = -\log f_{Cl^-} = [(0.51)(1)^2(1.00 \times 10^{-3})^{\frac{1}{2}}]/[1 + (1.00 \times 10^{-3})^{\frac{1}{2}}]$$

$$= 0.015_6$$

$$f_{Na^+} = f_{Cl^-} = 10^{-0.015_6} = 10^{-1} \times 10^{.984_4} = 0.96_5$$

17. $\mu = [(0.0040)(1)^2 + (0.0050)(2)^2 + (0.0020)(3)^2]/2 = 0.021$

This is >0.01, so use Equation 6.19. From Recommended Reference 9 in Chapter 6, $\alpha_{Na} = 4$, $\alpha_{Al} = 9$, $\alpha_{SO_4{}^{2-}} = 4$.

$$-\log f_{Na^+} = [(0.51)(1)^2(0.021)^{\frac{1}{2}}]/[1 + (0.33)(4)(0.021)^{\frac{1}{2}}] = 0.062$$

$$f_{Na^+} = 0.867$$

$$-\log f_{SO_4{}^{2-}} = [(0.51)(2)^2(0.021)^{\frac{1}{2}}]/[1 + (0.33)(4)(0.021)^{\frac{1}{2}}] = 0.25$$

$$f_{SO_4{}^{2-}} = 0.56$$

$$-\log f_{Al^{3+}} = [(0.51)(3)^2(0.021)^{\frac{1}{2}}]/[1 + (0.33)(9)(0.021)^{\frac{1}{2}}] = 0.46$$

$$f_{Al^{3+}} = 0.35$$

18. $\mu = [(0.0020)(1)^2 + (0.0020)(1)^2]/2 = 0.0020$

Since $\mu < 0.01$, use Equation 6.20:

$$-\log f_{NO_3{}^-} = [(0.51)(1)^2(0.0020)^{\frac{1}{2}}]/[1 + (0.0020)^{\frac{1}{2}}] = 0.022$$

$$f_{NO_3{}^-} = 0.95$$

$$a_{NO_3{}^-} = (0.0020)(0.95) = 0.0019\ \underline{M}$$

19. $\mu = [(0.040)(1)^2 + (0.020)(2)^2]/2 = 0.060$

Since $\mu > 0.01$, use Equation 6.19. From Recommended Reference 9 in Chapter 6, $\alpha_{CrO_4{}^{2-}} = 4$.

$$\log f_{CrO_4{}^{2-}} = [(0.51)(2)^2(0.060)^{\frac{1}{2}}]/[1 + (0.33)(4)(0.060)^{\frac{1}{2}}] = 0.37_8$$

$$f_{CrO_4{}^{2-}} = 0.42$$

$$a_{CrO_4{}^{2-}} = (0.020)(0.42) = 0.0084\ \underline{M}$$

20. (a) $K_a^\circ = (a_{H^+} \cdot a_{CN^-})/a_{HCN} = ([H^+]\, f_{H^+} \cdot [CN^-]\, f_{CN^-})/[HCN] = K_a\, f_{H^+} f_{CN^-}$

(b) $K_b^\circ = (a_{NH_4^+} \cdot a_{OH^-})/a_{NH_3} = ([NH_4^+]\, f_{NH_4^+} \cdot [OH^-]\, f_{OH^-})/[NH_3] = K_b f_{NH_4^+} f_{OH^-}$

21. (a) $HBenz = H^+ + Benz^-$

From Appendix C (constants at $\mu = 0$),

$K_a^\circ = [H^+][Benz^-]/[HBenz] = 6.3 \times 10^{-5}$

$(x)(x)/(5.0 \times 10^{-3}) = 6.3 \times 10^{-5}$ (Neglect x compared to C, which is $\approx$ $100 \times K_a^\circ$)

$x = 5.6 \times 10^{-4}$ $\underline{M} = [H^+]$

$pH = -\log(5.6 \times 10^{-4}) = 3.25$

(b) $\mu = [(0.100)(1)^2 + (0.050)(2)^2]/2 = 0.15$

From Recommended Reference 9 in Chapter 6, $\alpha_{H^+} = 9$.

$-\log f_{H^+} = [(0.51)(1)^2(0.15)^{\frac{1}{2}}]/[1 + (0.33)(9)(0.15)^{\frac{1}{2}}] = 0.092$

$f_{H^+} = 0.81$

From Recommended Reference 9 in Chapter 6, $\alpha_{Benz^-} = 6$.

$-\log f_{Benz^-} = [(0.51)(1)^2(0.15)^{\frac{1}{2}}]/[1 + (0.33)(6)(0.15)^{\frac{1}{2}}] = 0.11_1$

$f_{Benz^-} = 0.77$

At $\mu = 0.15$,

$K_a = K_a^\circ(f_{HA}/f_{H^+} f_{A^-}) = K_a^\circ(1/f_{H^+} f_{A^-})$

$K_a = (6.3 \times 10^{-5})/(0.81)(0.77) = 1.0_1 \times 10^{-4}$

$(x)(x)/(5.0 \times 10^{-3} - x) = 1.0_1 \times 10^{-4}$

Solving quadratic formula,

$x = 6.6 \times 10^{-4}$ $\underline{M} = [H^+]$

$pH = -\log(6.6 \times 10^{-4}) = 3.18$

22. *See the CD for the spreadsheet setup.*

23. *See the CD for the spreadsheet. The formula for f_i is:*

$$f_i = 10^\wedge\text{-}(0.51*Z_{i\ cell}{}^\wedge 2*\mu_{cell}{}^\wedge 0.5)/(1+\mu_{cell}{}^\wedge 0.5)$$

24. *See the CD for the spreadsheet. The spreadsheet calculated values are 0.919 for K^+ (vs. 0.918) and 0.713 for SO_4^{2-} (vs. (0.713). The slight difference is due to rounding in the manual calculation.*

25. *See the CD for the spreadsheets (a and b). The spreadsheet values are 0.794 for K^+ (vs. 0.792) and 0.419 for SO_4^{2-} (vs. 0.419).*

26. *See the CD for the spreadsheet calculations, Problems 16, 17, 18, and 19.*

CHAPTER 7

1. *A strong electrolyte is completely ionized in solution, while a weak electrolyte is only partially ionized in solution. A slightly soluble salt is generally a strong electrolyte, because it is completely ionized in solution.*

2. *The Brønsted acid-base theory assumes that an acid is a proton donor, and a base is a proton acceptor.*

3. *A conjugate acid is the protonated form of a Brønsted base, and a conjugate base is the ionized form of a Brønsted acid: Conjugate acid = H^+ + conjugate base.*

4. $C_6H_5NH_2 + HOAc \longrightarrow C_6H_5NH_3^+ + OAc^-$

   conjugate acid (under $C_6H_5NH_3^+$)

   $C_6H_5NH_2 + NH_2CH_2CH_2NH_2 \longrightarrow C_6H_5O^- + NH_2CH_2CH_2NH_3^+$

   conjugate base (under $C_6H_5O^-$)

5. *In the Lewis theory, an acid is an electron acceptor, while a base is an electron donor.*

6. (a) $pH = -\log 2.0 \times 10^{-2} = 2 - 0.30 = 1.70$

   $pOH = 14.00 - 1.70 = 12.30$

   (b) $pH = -\log 1.3 \times 10^{-4} = 4 - 0.11 = 3.89$

   $pOH = 14.00 - 3.89 = 10.11$

   (c) $pH = -\log 1.2 = -0.08$

   $pOH = 14.00 - (-0.08) = 14.08$

   (d) The concentration of HCl is about 100 times less than the concentration of H^+ from the ionization of water. Therefore, the former can be neglected and pH = pOH = 7.00. ($pH \neq -\log 1.2 \times 10^{-9} = 8.92$, which is alkaline!)

   (e) The contribution from water is appreciable. Therefore, use K_w to calculate its contribution.

$$H_2O = H^+ + OH^-$$

$$(2.4 \times 10^{-7} + x) \qquad x$$

$$(2.4 \times 10^{-7} + x)(x) = 1.0 \times 10^{-14}$$

From the quadratic equation, $x = 3._5 \times 10^{-8}$ $\underline{M}$

$[H^+] = 2.4 \times 10^{-7} + 0.4 \times 10^{-7} = 2.8 \times 10^{-7}$ $\underline{M}$

$pH = -\log 2.8 \times 10^{-7} = 7 - 0.45 = 6.55$ (slightly acid)

$pOH = 14.00 - 6.55 = 7.45$

7. (a) $pOH = -\log 5.0 \times 10^{-2} = 2 - 0.70 = 1.30$

$pH = 14.00 - 1.30 = 12.70$

(b) $pOH = -\log 2.8 \times 10^{-1} = 1 - 0.45 = 0.55$

$pH = 14.00 - 0.55 = 13.45$

(c) $pOH = -\log 2.4 = -0.38$

$pH = 14.00 - (-0.38) = 14.38$

(d) Must calculate contribution from H_2O

$H_2O = H^+ + OH^-$

$x \quad 3.0 \times 10^{-7} + x$

$(x)(3.0 \times 10^{-7} + x) = 1.0 \times 10^{-14}$

From the quadratic equation; $x = 2 \times 10^{-8}$ $\underline{M}$

$[OH^-] = 3.0 \times 10^{-7} \times 0.2 \times 10^{-7} = 3.2 \times 10^{-7}$ $\underline{M}$

$pOH = -\log 3.2 \times 10^{-7} = 7 - 0.51 = 6.49$

$pH = 14.00 - 6.49 = 7.51$

(e) $pOH = -\log 3.7 \times 10^{-3} = 3 - 0.57 = 2.43$

$pH = 14.00 - 2.43 = 11.57$

8. (a) $[OH^-] = (1.0 \times 10^{-14})/(2.6 \times 10^{-5}) = 3.8 \times 10^{-10}$ $\underline{M}$

(b) $[OH^-] = (1.0 \times 10^{-14})/(0.20) = 5.0 \times 10^{-14}$ $\underline{M}$

(c) $[H^+] = 1.0 \times 10^{-7}$ $\underline{M}$ ($HClO_4$ is negligible)

$[OH^-] = 1.0 \times 10^{-7}$ $\underline{M}$

(d) $[OH^-] = (1.0 \times 10^{-14})/(1.9) = 5.3 \times 10^{-15}$ $\underline{M}$

9. (a) $[H^+] = 10^{-3.47} = 10^{-4} \times 10^{.53} = 3.4 \times 10^{-4}$ $\underline{M}$

(b) $[H^+] = 10^{-.20} = 10^{-1} \times 10^{.80} = 6.3 \times 10^{-1}$ $\underline{M}$ (0.63 $\underline{M}$)

(c) $[H^+] = 10^{-8.60} = 10^{-9} \times 10^{.40} = 2.5 \times 10^{-9}$ $\underline{M}$

(d) $[H^+] = 10^{-(-.60)} = 10^{.60} = 4.0$ $\underline{M}$

(e) $[H^+] = 10^{-14.35} = 10^{-15} \times 10^{.65} = 4.5 \times 10^{-15}$ $\underline{M}$

(f) $[H^+] = 10^{1.25} = 10^1 \times 10^{.25} = 1.8 \times 10^1$ $\underline{M}$ (18 $\underline{M}$)

10. Assume 1.0 mL of each is mixed.

Excess NaOH = (0.30 $\underline{M}$ x 1.0 mL - 0.10 $\underline{M}$ x 1.0 mL x 2)/2 mL

= 0.10 mmol/2 mL = 0.050 $\underline{M}$

$pOH = -\log 5.0 \times 10^{-2} = 2 - 0.70 = 1.30$

$pH = 14.00 - 1.30 = 12.70$

11. Assume 1.0 mL volumes.

$[H^+]$ of acid solution = 1.0×10^{-3} $\underline{M}$

$[H^+]$ of base solution = 1.0×10^{-12} $\underline{M}$;

$[OH^-] = (1.0 \times 10^{-14})/(1.0 \times 10^{-12}) = 1.0 \times 10^{-2}$ $\underline{M}$

Excess base = (1.0×10^{-2} $\underline{M}$ x 1.0 mL - 1.0×10^{-3} $\underline{M}$ x 1.0 mL)/2 mL

= 9×10^{-3} mmol/2 mL = 4.5×10^{-3} $\underline{M}$ OH^-

$\therefore$ $pOH = -\log 4.5 \times 10^{-3} = 3 - 0.65 = 2.35$

$pH = 14.00 - 2.35 = 11.65$

12. $[H^+][OH^-] = 5.5 \times 10^{-14}$

$[H^+]^2 = 5.5 \times 10^{-14}$

$[H^+] = 2.3 \times 10^{-7}$ $\underline{M}$

$pH = -\log 2.3 \times 10^{-7} = 7 - 0.36 = 6.64$

13. $pH + pOH = 13.60$

$pOH = 13.60 - 7.40 = 6.20$

14. $[H^+] = [OAc^-] = 10^{-3.26} = 10^{.74} \times 10^{-4} = 5.5 \times 10^{-4}$ $\underline{M}$

$(5.5 \times 10^{-4})^2/[HOAc] = 1.75 \times 10^{-5}$

$[HOAc] = 1.7_3 \times 10^{-2}$ $\underline{M}$ (neglecting $[H^+]$ in the denominator).

% ionized = $[(5.5 \times 10^{-4})/(1.7_3 \times 10^{-2})] \times 100\% = 3.2\%$

15. $pOH = 14.00 - 8.42 = 5.58$

$[OH^-] = [RNH_3^+] = 10^{-5.58} = 10^{.42} \times 10^{-6} = 2.6 \times 10^{-6}$ $\underline{M}$

$K_b = (2.6 \times 10^{-6})^2/(0.20) = 3.4 \times 10^{-11}$

$pK_b = -\log 3.4 \times 10^{-11} = 10.47$

16. Let x = concentration of acid

$(0.035x)^2/(x) = 6.7 \times 10^{-4}$

$x = 0.55$ $\underline{M}$

100 g/L = 0.55 mol/L

∴ f.w. = 100 g/0.55 mol = 18_2 g/mol

17. $[H^+][Prop^-]/[H\ Prop] = 1.3 \times 10^{-5}$

$(x)(x)/(0.25) = 1.3 \times 10^{-5}$

$x = 1.8 \times 10^{-3}$

$pH = -\log 1.8 \times 10^{-3} = -(3 - 0.26) = 2.74$

18. $RNH_2 + H_2O = RNH_3^+ + OH^-$

$K_b = [RNH_3^+][OH^-]/[RNH_2]$

$4.0 \times 10^{-10} = (x)(x)/(0.10)$

$x = 6.3 \times 10^{-6}$

$pOH = -\log 6.3 \times 10^{-6} = -(6 - 0.80) = 5.20$

$pH = 14.00 - 5.20 = 8.80$

19. $[H^+][IO_3^-]/[HIO_3] = 2 \times 10^{-1}$

$HIO_3 \quad = H^+ + IO_3^-$

$0.1 - x \quad x \quad x$

$(x)(x)/(0.1 - x) = 2 \times 10^{-1}$

Since K_a is large, the quadratic equation must be solved:

$x^2 = 0.2 \times 0.1 - 0.2x$

$x^2 + 0.2x - 0.02 = 0$

$x = [-0.2 \pm \sqrt{(0.2)^2 - (4)(1)(0.02)}]/(2 \times 1)$

$x = 0.07$ M $= [H^+]$

pH = $-\log 0.07 = -(2 - 0.8) = 1.2$

20. $H_2SO_4 \rightarrow H^+ + HSO_4^-$

$HSO_4^- = H^+ + SO_4^{2-}$

The total $[H^+] = [HSO_4^-] + [SO_4^{2-}]$. Let $x = [SO_4^{2-}]$. Then

$[HSO_4^-] = 0.0100 - x$

$[H^+] = 0.0100 + x$

$K_{a2} = [H^+][SO_4^{2-}]/[HSO_4^-] = 1.2 \times 10^{-2}$

K_{a2} is not very small, so x is probably appreciable and can't be neglected.

$\therefore$ $(0.0100 + x)(x)/(0.0100 - x) = 1.2 \times 10^{-2}$

The quadratic formula must be used to for x:

$x^2 + 0.022x - 1.2 \times 10^{-4} = 0$

$x = [-0.022 \pm \sqrt{(0.022)^2 - 4(-1.2 \times 10^{-4})}]/2 = 0.004_5$ M

$[H^+] = 0.0100 + 0.004_5 = 0.014_5$ M

21. From the Appendix, $K_a = 0.129$

HT =	H^+ +	T^-
0.100-x	x	x

The initial concentration is much less than 200 K_a, so the quadratic equation must be solved.

$[H^+][T^-]/[HT] = 0.129$

$(x)(x)/(0.100-x) = 0.129$

$x^2 + 0.129x - 0.0129 = 0$

$x = [-0.129 \pm \sqrt{(0.129)^2 - 4(-0.0129)}]/2 = 0.0661\ \underline{M}$

22. $RNH_2 + H_2O = RNH_3^+ + OH^-$

$0.20-x \quad x \quad x$

$K_b = 10^{-4.20} = 10^{-5} \times 10^{.80} = 6.3 \times 10^{-5}$

$[RNH_3^+][OH^-]/[RNH_2] = 6.3 \times 10^{-5}$

$(x)(x)/(0.20) = 6.3 \times 10^{-5}$

$x = 3.5 \times 10^{-3}\ \underline{M} = [OH^-]$

$pOH = -\log 3.5 \times 10^{-3} = -(-3 + 0.54) = 2.46$

$pH = 14.00 - 2.46 = 11.54$

23. $HOAc = H^+ + OAc^-$

$c-0.030c \quad 0.030c \quad 0.030c$

$\approx c$

$(0.030c)(0.030c)/c = 1.75 \times 10^{-5}$

$9.0 \times 10^{-4}\ c^2 = 1.75 \times 10^{-5}c$

$c = 0.019\ \underline{M}$

24. $HA = H^+ + A^-$

$c-x \quad x \quad x$

$\approx c$ (since $c > 200\ K_a$)

Assume 1% ionized at 0.100 $\underline{M}$. Then $x = 0.00100$

$(0.00100)^2/(0.100) = K_a = 1.00 \times 10^{-5}$

For 2% ionized (doubled), $x = 0.0200c$

$(0.02c)^2/c = 1.00 \times 10^{-5}$

$4.00 \times 10^{-4}c^2 = 1.00 \times 10^{-5}c$

$c = 0.025\ \underline{M}$

Therefore, must dilute HA 4-fold.

25. $H_3BO_3 + NaOH \rightarrow NaBO_2 + 2H_2O$

mmol H_3BO_3 = 20 mL × 0.25 $\underline{M}$ = 5.0 mmol

mmol NaOH = 25 mL × 0.20 $\underline{M}$ = 5.0 mmol

These are stoichiometrically equal, so 5.0 mmol BO_2^- is formed in 45 mL.

$[BO_2^-] = 5.0\text{ mmol}/45\text{ mL} = 0.11_1\text{ mmol/mL}$

$BO_2^- + 2H_2O = H_3BO_3 + OH^-$

$0.11_1 - x \qquad x \qquad x$

$[H_3BO_3][OH^-]/[BO_2^-] = K_b = K_w/K_a = (1.0 \times 10^{-14})/(6.4 \times 10^{-10}) = 1.6 \times 10^{-5}$

$c > 200\ K_b$. $\therefore\ c \approx 0.11_1\ \underline{M}$

$(x)(x)/(0.11_1) = 1.6 \times 10^{-5}$

$x = 1.3_3 \times 10^{-3}\ \underline{M} = [OH^-]$

Or, from Equation 7.32,

$[OH^-] = \sqrt{K_b \cdot c_{A^-}} = \sqrt{(1.6 \times 10^{-5})(0.11_1)} = 1.3_3 \times 10^{-3}\ \underline{M}$

pOH = 2.88; pH = 14.00 − 2.88 = 11.12

26. $CN^- + H_2O = HCN + OH^-$

$0.010 - x \qquad x \qquad x$

$[HCN][OH^-]/[CN^-] = K_b = K_w/K_a = 1.0 \times 10^{-14}/7.2 \times 10^{-10} = 1.4 \times 10^{-5}$

$C_{CN^-} > 100\ K_b$. $\therefore$ neglect x compared to 0.010 $\underline{M}$ and use Equation 7.32.

$[OH^-] = \sqrt{K_b \cdot C_{A^-}} = \sqrt{(1.4 \times 10^{-5})(0.010)} = 3.7 \times 10^{-4}\ \underline{M}$

$pOH = 3.43;\ pH = 14.00 - 3.43 = 10.57$

27. K_a benzoic acid $= 6.3 \times 10^{-5}$

$C_6H_5COO^- + H_2O = C_6H_5COOH + OH^-$

From Equation 7.32 (since $C_{A^-} > 200\ K_b$),

$[OH^-] = \sqrt{(K_w/K_a[C_6H_5COO^-]}$

$= \sqrt{[(1.0 \times 10^{-14})/(6.3 \times 10^{-5})] \times 0.050} = 2.8 \times 10^{-6}\ \underline{M}$

$pOH = -\log 2.8 \times 10^{-6} = 5.55$

$pH = 14.00 - 5.55 = 8.45$

28. K_b pyridine $= 1.7 \times 10^{-9}$

$C_6H_5NH^+ + H_2O = C_6H_5NHOH + H^+$; $C_{BH^+} > 100\ K_a$, so use Equation 7.39 .

$[H^+] = \sqrt{(K_w/K_b)[C_6H_5NH^+]} = \sqrt{[(1.0 \times 10^{-14})/(1.7 \times 10^{-9})] \times 0.25} = 1.2 \times 10^{-3}\underline{M}$

$pH = -\log 1.2 \times 10^{-3} = 2.92$

29. mmol $H^+ = 0.25\ \underline{M} \times 12.0\ mL \times 2 = 6.0$ mmol

mmol $NH_3 = 1.0\ \underline{M} \times 6.0\ mL = 6.0$ mmol

∴ 6.0 mol of NH_4^+ are formed in a volume of 18.0 mL.

$[NH_4^+] = 6.0\ mmol/18\ mL = 0.33\ \underline{M}$

$[H^+] = \sqrt{[(1.0 \times 10^{-14})/(1.75 \times 10^{-5})] \times (0.33)} = 1.4 \times 10^{-5}\ \underline{M}$

$pH = 5 - \log 1.5 = 5 - 0.15 = 4.85$

30. mmol HOAc $= 0.10\ \underline{M} \times 20\ mL = 2.0$ mmol

mmol NaOH $= 0.10\ \underline{M} \times 20\ mL = 2.0$ mmol

∴ 2.0 mmol of NaOAc are formed in a volume of 40 mL.

$[OAc^-] = 2.0 \text{ mmol}/40 \text{ mL} = 0.050\ \underline{M}$

$[OH^-] = \sqrt{[(1.0 \times 10^{-14})/1.75 \times 10^{-5})] \times (0.050)} = 5.3 \times 10^{-6}\ \underline{M}$

$pOH = 6 - \log 5.3 = 6 - 0.72 = 5.28$

$pH = 14.00 - 5.28 = 8.72$

31. $HONH_2 + HCl = HONH_3^+ + Cl^-$

We form 0.10 mol $HONH_3^+/0.50\ L = 0.20\ \underline{M}$

$HONH_3^+ + H_2O = HONH_3OH + H^+$

$[HONH_3OH][H^+]/[HONH_3^+] = K_a = K_w/K_b = 1.0 \times 10^{-14}/9.1 \times 10^{-9} = 1.1 \times 10^{-6}$

$C_{BH^+} > 200\ K_a$, therefore,

$[H^+] = \sqrt{K_a \cdot C_{BH^+}} = \sqrt{(1.1 \times 10^{-6})(0.20)} = 4.7 \times 10^{-4}\ \underline{M}$

$pH = 3.33$

32. $C_6H_5(OH)COO^- + H_2O = C_6H_5(OH)COOH + OH^-$

$c-x \qquad x \qquad x$

$[C_6H_5(OH)COOH][OH^-]/[C_6H_5(OH)COO^-] = K_b = K_w/K_a = 1.0 \times 10^{-14}/1.0 \times 10^{-3}$

$= 1.0 \times 10^{-11}$

$C_{A^-} > 200\ K_b$, therefore,

$[OH^-] = \sqrt{(1.0 \times 10^{-11})(0.0010)} = 1.0 \times 10^{-7}\ \underline{M}$

$pOH = pH = 7.00$

(Salicylic acid is strong enough that at dilute concentrations of its salt, the extent of hydrolysis is negligible.)

33. $CN^- + H_2O = HCN + OH^-$

$1.0 \times 10^{-4} - x \qquad x \qquad x$

From Problem 26, $K_b = 1.4 \times 10^{-5}$. $C_{A^-} < 200\ K_b$, so we can't neglect x.

Therefore, solve the quadratic formula. This gives for $(x)(x)/[1.0 \times 10^{-4} - x] = 1.4 \times 10^{-5}$:

$x = [OH^-] = 3.1 \times 10^{-5}$ $\underline{M}$ (the hydrolysis is 31% complete)

pOH = 4.51; pH = 9.49

34. $H_2P \;=\; H^+ + HP^-$

$0.0100-x \quad x \quad x$

Assume second ionization is small.

$[H^+][HP^-]/[H_2P] = K_{a_1}$

$(x)(x)/(0.0100-x) = 1.2 \times 10^{-3}$

$c < 200\,K_a$. Therefore must solve by quadratic equation.

$x = [-1.2 \times 10^{-3} \pm \sqrt{(1.2 \times 10^{-3})^2 - 4(-1.2 \times 10^{-5})}]/2 = 2.9 \times 10^{-3}$ $\underline{M}$

$pH = -\log 2.9 \times 10^{-3} = 2.54$

35. $P^{2-} + H_2O = HP^- + OH^-$

$0.0100-x \quad x \quad x$

Assume further hydrolysis of HP^- is negligible.

$[HP^-][OH^-]/[P^{2-}] = K_b = (K_w/K_{a_2}) = (1.0 \times 10^{-14})/(3.9 \times 10^{-6}) = 2.6 \times 10^{-9}$

K_b is small, so x can be neglected compared to 0.0100 $\underline{M}$, and the quadratic equation need not be used.

$[OH^-] = \sqrt{(K_w/K_{a_2}) \times [P^{2-}]} = \sqrt{2.6 \times 10^{-9} \times 1.00 \times 10^{-2}} = 5.1 \times 10^{-6}$ $\underline{M}$

pOH = 5.29; pH = 14.00 - 5.29 = 8.71

36. This is an amphoteric salt:

$HP^- = H^+ + P^{2-} \qquad K_{a_2}$

$HP^- + H_2O = H_2P + OH^- \qquad K_b = K_w/K_{a_1}$

$[H^+] = \sqrt{K_{a_1}K_{a_2}} = \sqrt{1.2 \times 10^{-3} \times 3.9 \times 10^{-6}} = 6.9 \times 10^{-5}$ M

pH = $-\log 6.9 \times 10^{-5} = 5 - 0.84 = 4.16$

37. S^{2-} hydrolyzes:

$S^{2-} + H_2O = HS^- + OH^-$

0.600-x x x

$K_b = (K_w/K_{a_2}) = (1.0 \times 10^{-14})/(1.2 \times 10^{-15}) = 8.3$

The equilibrium therefore lies significantly to the right and the quadratic equation must be used to solve for the degree of ionization:

$[HS^-][OH^-]/[S^{2-}] = (x)(x)/(0.600-x) = 8.3$

$x^2 + 8.3x - 4.9_8 = 0$

$x = [-8.3 \pm \sqrt{(8.3)^2 - 4(-4.9_8)}]/2 = 0.55$ M (92% hydrolyzed! Strong base)

pOH = $-\log 0.55 = 0.26$; pH = $14.00 - 0.26 = 13.74$

38. PO_4^{3-} hydrolyzes. From Equation 7.87, $K_b = (K_w/K_{a_3}) = [HPO_4^{2-}][OH^-]/[PO_4^{3-}]$

$= (1.0 \times 10^{-14})/(4.8 \times 10^{-13}) = 0.021$

Since the equilibrium constant is not very small, the quadratic equation must be used to solve for the degree of ionization. If

$x = [OH^-] = [HPO_4^{2-}]$, and $[PO_4^{3-}] = 0.500-x$, then:

$(x^2)/(0.500-x) = 0.021$

$x^2 + 0.021x - 0.010_5 = 0$

$x = [-0.021 \pm \sqrt{(0.021)^2 - 4(-0.010_5)}]/2 = 0.092$ M

pOH = $-\log 0.092 = 1.04$; pH = $14.00 - 1.04 = 12.96$

39. HCO_3^- is amphoteric. For H_2CO_3, $K_{a_1} = 4.3 = 10^{-7}$ and $K_{a_2} = 4.8 \times 10^{-11}$.

Since the difference between these is large, we can write for HCO_3^-:

$[H^+] = \sqrt{K_{a_1}K_{a_2}} = \sqrt{4.3 \times 10^{-7} \times 4.8 \times 10^{-11}} = 4.5 \times 10^{-9}$ M

$pH = -\log 4.5 \times 10^{-9} = 8.34$

40. HS^- is amphoteric.

$[H^+] = \sqrt{K_{a_1} K_{a_2}} = \sqrt{9.1 \times 10^{-8} \times 1.2 \times 10^{-15}} = 1.0_4 \times 10^{-11}$ $\underline{M}$

$pH = -\log 1.04 \times 10^{-11} = 10.98$

41. $HY^{3-} = H^+ + Y^{4-}$ $\qquad K_{a_4} = 5.5 \times 10^{-11}$

$HY^{3-} + H_2O = H_2Y^{2-} + OH^-$ $\qquad K_b = K_w/K_{a_3} = 1.0 \times 10^{-14}/6.9 \times 10^{-7}$

$= 1.4 \times 10^{-8}$ (use K_a of the conjugate acid, H_2Y^{2-})

The difference between K_{a_3} and K_{a_4} is large, so

$[H^+] = \sqrt{K_{a_3} K_{a_4}} = \sqrt{(5.5 \times 10^{-11})(1.4 \times 10^{-8})} = 8.8 \times 10^{-10}$ $\underline{M}$

$pH = 9.06$

42. $pH = pKa + \log ([COO^-]/[HCOOH])$; $pKa = 3.75$

$pH = 3.75 + \log (0.10/0.050) = 3.75 + 0.30 = 4.05$

43. mmol $NH_3 = 0.10 \times 5.0 = 0.50$ mmol

mmol $HCl = 0.020 \times 10.0 = 0.20$ mmol

mmol NH_4^+ formed $= 0.20$ mmol

mmol excess $NH_3 = 0.50 - 0.20 = 0.30$ mmol

$pH = pKa + \log$ [proton acceptor]/[proton donor] $= (pK_w - pK_b) + \log [NH_3]/[NH_4^+]$

$= (14.00 - 4.76) + \log (0.30)/(0.20) = 9.24 + 0.18 = 9.42$

Or, from Equation 7.58,

$pOH = 4.76 + \log ([NH_4^+]/[NH_3] = 4.76 + \log (0.20.0.30) = 4.58$

$pH = 14.00 - 4.58 = 9.42$

44. $5.00 = 4.76 + \log (0.100/[HOAc])$

$[HOAc] = 0.058$ $\underline{M}$

mmol HOAc = 0.058 $\underline{M}$ x 100 mL = 5.8 mmol

mmol NaOAc = 0.100 $\underline{M}$ x 100 mL = 10.0 mmol

mmol NaOH added = 0.10 $\underline{M}$ x 10 mL = 1.0 mmol

After adding NaOH:

mmol NaOAc = 10.0 + 1.0 = 11.0

mmol HOAc = 5.8 - 1.0 = 4.8

pH = 4.76 + log (11.0/4.8) = 5.12

The pH increases by 5.12 - 5.00 = 0.12

45. mmol HOAc = 50 mL x 0.10 $\underline{M}$ = 5.0 mmol

mmol NaOH = 20 mL x 0.10 $\underline{M}$ = $\underline{2.0 \text{ mmol}}$ = mmol HOAc converted to OAc^-

mmol HOAc left = 3.0 mmol

$pH = pK_a + \log([OAc^-]/[HOAc]) = 4.76 + \log(2.0/3.0) = 4.58$

46. mmol NH_3 = 50 mL x 0.10 $\underline{M}$ = 5.0 mmol

mmol H_2SO_4 = 25 mL x 0.050 $\underline{M}$ = 1.2_5 mmol x 2 = 2.5 mmol H^+ = mmol NH_3 converted to NH_4^+

mmol NH_3 left = 2.5 mmol

$pK_a = pK_w - pK_b = 14.00 - 4.76 = 9.24$

$pH = pK_a + \log [NH_3]/[NH_4^+] = 9.24 + \log (2.5)/(2.5) = 9.24$

47. From the pH of the stomach contents and the pK_a of aspirin, we can calculate the mole ratio of the ionized to un-ionized forms:

$pH = pK_a + \log(\text{mmol } A^-/\text{mmol } HA)$

$2.95 = 3.50 + \log(\text{mmol } A^-/\text{mmol } HA)$

$(\text{mmol } A^-/\text{mmol } HA) = 0.28$; mmol A^- = mmol HA x 0.28

The total mmol of each form is 650 mg/180 (mg/mmol) = 3.6_1 mmol

mmol HA + mmol A^- = 3.6_1

mmol HA + 0.28 x mmol HA = 3.6_1

1.28 x mmol HA = 3.6_1

mmol HA = 2.8_2

mg HA = 2.8_2 mmol x 180 mg/mmol = 50_8 HA available for absorption.

48. $pK_a = pK_w - pK_b$

$pH = pK_a + \log([(HOCH_2)_3NH_2]/[(HOCH_2)_3NH_3^+])$

$7.40 = 8.08 + \log([(HOCH_2)_3NH_2]/[(HOCH_2)_3NH_3^+])$

$([(HOCH_2)_3NH_2]/[(HOCH_2)_3NH_3^+]) = 0.21 = ([THAM]/[THAMH^+])$

mmol HCl = 0.50 <u>M</u> x 100 mL = 50 mmol

.˙. must add enough THAM to form 50 mmol $THAMH^+$, plus enough excess to satisfy the above ratio.

Total THAM = 50 mmol + x mmol.

(x mmol/50 mmol) = 0.21

x = $10._5$ mmol THAM

Total THAM = 50 + 10 = 60 mmol

mg THAM = 60 mmol x 121.1 mg/mmol = 7,300 mg = 7.3 g

49. In this case, K_a of the acid is large, and the presence of the added salt, T^-, will not suppress ionization sufficiently to make x negligible.

HT	=	H^+	+	T^-
0.100-x		x		0.100+x

$(x)(0.100+x)/(0.100-x) = 0.129$

$x^2 + 0.229x - 0.0129 = 0$

$x = [-0.229 \pm \sqrt{(0.229)^2 - 4(-0.0129)}]/2 = 0.46_7$ M

This compares with 0.129 M ($[H^+] = K_a$) had x been assumed negligible.

50. $pH = pK_{a_1} + \log([HP^-]/[H_2P])$

$K_{a_1} = 1.2 \times 10^{-3}$; $pK_a = 2.92$

$pH = 2.92 + \log(0.10/0.20) = 2.92 - 0.30 = 2.62$

51. $pH = pK_{a_2} + \log([P^{2-}]/[HP^-])$

Since $[P^{2-}] = [HP^-]$,

$pH = pK_{a_2} = -\log 3.9 \times 10^{-6} = 5.41$

52. $pH = pK_2 + \log([HPO_4^{2-}]/[H_2PO_4^-])$

$7.45 = 7.12 + \log([HPO_4^{2-}]/[H_2PO_4^-])$

$[HPO_4^{2-}] + [H_2PO_4^-] = 3.0 \times 10^{-3}$

Solution of the two simultaneous equations gives:

$[HPO_4^{2-}] = 2.0 \times 10^{-3}$ M

$[H_2PO_4^-] = 9.6 \times 10^{-4}$ M

53. $\beta = 2.303\, C_{HA}C_{A^-}/(C_{HA} + C_{A^-})$

$= 2.303 \times 0.10 \times 0.070/(0.10 + 0.070)$

$= 0.095$ mol/L per pH (or 10.5 pH per mol/L acid or base)

The volume change can be neglected when adding 10 mL acid or base.

The HCl and NaOH are diluted 1000-fold, to 0.0010 M.

For 10 mL 0.10 M HCl:

$dpH = -dC_{HA}/\beta = -0.0010$ M/0.095 M per pH $= -0.011$ pH $= \Delta pH$

For 10 mL 0.10 M NaOH,

$dpH = dC_{BOH}/\beta = 0.0010$ M/0.095 M per pH $= +0.011$ pH $= \Delta pH$

54. At pH 4.76 (=pK_a), [HOAc] = [OAc^-] = x

$\beta = 2.303\, C_{HA}C_{A^-}/(C_{HA} + C_{A^-})$

1.0 M per pH = $2.303x^2/(2x)$

x = 0.868 M = [HOAc] = [OAc^-]

55. Let x = M Na_2HPO_4, y = M KH_2PO_4

We have two unknowns and need two equations.

$\mu = 1/2 \sum C_i Z_i^2$

$0.20 = 1/2\ ([Na^+](1)^2 + [HPO_4^{2-}](2)^2 + [K^+](1)^2 + [H_2PO_4^-](1)^2)$

$0.20 = 1/2[2x(1)^2 + x(2)^2 + y(1)^2 + y(1)^2]$

$0.20 = 3x + y$

$pH = pK_2 + \log\ ([HPO_4^{2-}]/[H_2PO_4^-])$

$7.40 = 7.12 + \log\ (x/y)$

Solution of the two simultaneous equations for x and y gives:

$x = 0.057\ \underline{M}\ Na_2HPO_4$

$y = 0.030\ \underline{M}\ KH_2PO_4$

mg $Na_2HPO_4 = 0.057\ \underline{M} \times 200$ mL $\times$ 142 mg/mmol = 1,620 mg = 1.6_2g

mg $KH_2PO_4 = 0.030\ \underline{M} \times 200$ mL $\times$ 136 mg/mmol = 820 mg = 0.82 g

56. Let $x = \underline{M}\ KH_2PO_4$, $y = \underline{M}\ H_3PO_4$

H_3PO_4 is non-ionic and does not contribute to the ionic strength.

$\mu = 1/2 \sum C_iZ_i^2 = 1/2\ ([K^+](1)^2 + [H_2PO_4^-](1)^2)$

$\therefore\ 0.20 = 1/2(x(1)^2 + x(1)^2) = x$

$KH_2PO_4 = 0.20\ \underline{M}$

mg $KH_2PO_4 = 0.20\ \underline{M} \times 200$ mL $\times$ 136 mg/mmol = 5,400 mg = 5.4 g

$pH = pK_1 + \log\ ([KH_2PO_4]/[H_3PO_4])$

$3.00 = 1.96 + \log\ (0.20/y)$

$y = 0.018\ \underline{M} = 0.018\ \underline{M}\ H_3PO_4$

$\underline{M}$ conc. H_3PO_4 = (0.85 g/g soln $\times$ 1.69 g soln/mL)/(98.0 g/mol)

= 0.015 mol/mL = 15 mmol/mL

15 $\underline{M}$ $\times$ mL = 0.018 $\underline{M}$ $\times$ 200 mL

mL conc. H_3PO_4 = 0.24 mL

57. From the Appendix, $K_{a_1} = 1.3 \times 10^{-2}$, $K_{a_2} = 5 \times 10^{-6}$. The equilibria are

$H_2SO_3 = H^+ + HSO_3^-$ (1)

$HSO_3^- = H^+ + SO_3^{2-}$ (2)

The total concentration of sulfurous acid, $C_{H_2SO_3}$, is given by

$C_{H_2SO_3} = [SO_3^{2-}] + [HSO_3^- + [H_2SO_3]$ (3)

Using the equilibrium constant expressions to solve for the various sulfurous acid species in terms of $[H_2SO_3]$ and substituting in (3), we reach

$C_{H_2SO_3} = (K_{a_1}K_{a_2}[H_2SO_3])/[H^+]^2) + (K_{a_1}[H_2SO_3])/([H^+]) + [H_2SO_3]$ (4)

from which

$1/\alpha_0 = [C_{H_2SO_3}]/[H_2SO_3] = K_{a_1}K_{a_2}/[H^+]^2 + K_{a_1}K_{a_2}/[H^+] + 1 = 138$

(at pH 4.00) (5)

Similarly,

$1/\alpha_1 = [C_{H_2SO_3}]/[HSO_3^-] = K_{a_2}/[H^+] + 1 + [H^+]/K_{a_1} = 1.06$ (6)

$1/\alpha_2 = [C_{H_2SO_3}]/[SO_3^{2-}] = 1 + [H^+]/K_{a_2} + [H^+]^2/(K_{a_1}K_{a_2}) = 21$ (7)

$[H_2SO_3] = C_{H_2SO_3}/(1/\alpha_0) = 0.0100/138 = 7.24 \times 10^{-5}$ $\underline{M}$

$[HSO_3^-] = C_{H_2SO_3}/(1/\alpha_1) = 0.0100/1.06 = 9.43 \times 10^{-3}$ $\underline{M}$

$[SO_3^{2-}] = C_{H_2SO_3}/(1/\alpha_2) = 0.0100/21 = 4.8 \times 10^{-4}$ $\underline{M}$

58. $C_{H_3PO_4} = [H_3PO_4] + [H_2PO_4^-] + [HPO_4^{2-}] + [PO_4^{3-}]$ (1)

Solving first for α_3, substitute in (1) for the various concentrations in terms of $[PO_4^{3-}]$. From the equilibrium constant expressions,

$[HPO_4^{2-}] = ([H^+][PO_4^{3-}])/K_{a_3}$ (2)

$[H_2PO_4^-] = ([H^+][HPO_4^{2-}])/(K_{a_2}) = ([H^+]^2[PO_4^{3-}])/(K_{a_2}K_{a_3})$ (3)

$[H_3PO_4] = ([H^+][H_2PO_4^-])/(K_{a_1}) = ([H^+]^3[PO_4^{3-}])/(K_{a_1}K_{a_2}K_{a_3})$ (4)

Substituting in (1):

$C_{H_3PO_4} = ([H^+]^3[PO_4^{3-}])/(K_{a_1}K_{a_2}K_{a_3}) + ([H^+]^2[PO_4^{3-}])/(K_{a_2}K_{a_3})$

$+ ([H^+][PO_4^{3-}])/(K_{a_3}) + [PO_4^{3-}]$ (5)

which, multiplying by $(K_{a_1}K_{a_2}K_{a_3}/K_{a_1}K_{a_2}K_{a_3})$ is

$C_{H_3PO_4} = ([H^+]^3[PO_4^{3-}] + K_{a_1}[H^+]^2[PO_4^{3-}] + K_{a_1}K_{a_2}[H^+][PO_4^{3-}]$

$+ K_{a_1}K_{a_2}K_{a_3}[PO_4^{3-}])/(K_{a_1}K_{a_2}K_{a_3})$ (6)

Substituting this in the denominators of the α_3 expression:

$\alpha_3 = [PO_4^{3-}]/[C_{H_3PO_4}] = (K_{a_1}K_{a_2}K_{a_3})/[H^+]^3 + K_{a_1}[H^+]^2 + K_{a_1}K_{a_2}[H^+]$

$+ K_{a_1}K_{a_2}K_{a_3})$ (7)

We can take a similar approach for α_1 and α_2, or else we can use (6) as the denominator for α_1 and α_2 and substitute (3) or (2), respectively, for the numerator. The result gives Equations 7.73 and 7.74.

59. $\mu = 0.100$; $f_{H^+} = 0.83$; $f_{CN^-} = 0.76$; $K_a° = 7.2 \times 10^{-10}$

$HCN = H^+ + CN^-$

0.0200-x x x

≈ 0.0200

$K_a° = ([H^+][CN^-]\, f_{H^+}\, f_{CN^-})/([HCN])$

$7.2 \times 10^{-10} = [(x)(x)(0.83)(0.76)]/(0.0200)$

$x = [H^+] = 4.8 \times 10^{-6}$ <u>M</u>

60. $B + H_2O = BH^+ + OH^-$

$K_b° = a_{BH^+} \cdot a_{OH^-}/a_B \approx a_{BH^+} \cdot a_{OH^-}/[B]$

$K_b° = [BH^+]\, f_{BH^+} \cdot [OH^-]\, f_{OH^-}/[B] = K_b f_{BH^+}\, f_{OH^-}$

$K_b = K_b°/f_{BH^+}f_{OH^-}$

61. *The system point is pH = 4.76 at 10^{-3} M HOAc.*

This is used as the reference point for the slope = -1 or +1 plots for log [HOAc] or log [OAc$^-$]. In strongly acid solution, [HOAc] ≈ 10^{-3} M and in strongly alkaline solution, [OAc$^-$] ≈ 10^{-3} M. At pH = pK_a = 4.76, [HOAc] = [OAc$^-$] = 5 x 10^{-4} M, and log [HOAc] = log [OAc$^-$] = -3.30. The log [H$^+$] and log [OH$^-$] curves are as in Figure 7.2 (or 7.3). Using the above data to plot the curves gives:

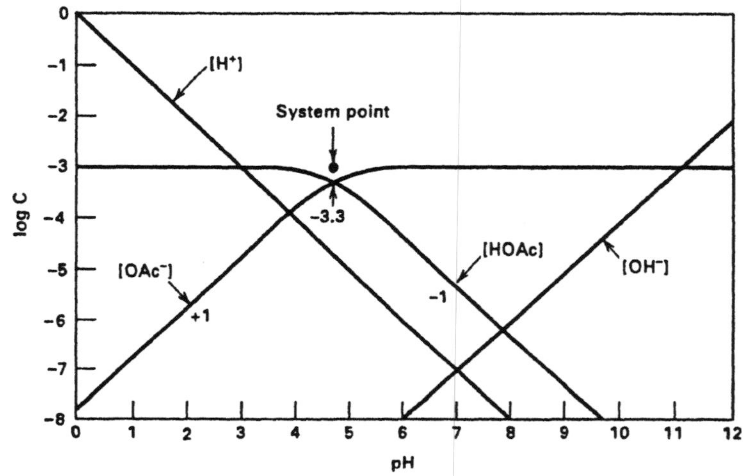

See the CD for the spreadsheet setup and chart of the log-log curves.

62. *In acid solution,* $[H^+] \approx [OAc^-]$. *From the plot where the* $[H^+]$ *and* $[OAc^-]$ *curves cross, log* $[H^+] = -3.88$, $pH = 3.88$. *Likewise,* $[OAc^-] = 10^{-3.88}$ $\underline{M}$.

63. $\log [OAc^-] = \log C_{HOAc}K_a + pH$

$\log [OAc^-] = \log (10^{-3})(1.75 \times 10^{-5}) + pH$

$\log [OAc^-] = -7.76 + pH$

For pH 2.00:

$\log [OAc^-] = -7.76 + 2.00 = -5.76$

$[OAc^-] = 10^{-5.76} = 1.7 \times 10^{-6}$ $\underline{M}$

A similar value is estimated from the log-log diagram.

64. *See CD, Chapter 7 Problems for spreadsheet construction of malic acid log-log diagram using alpha values.*

65. (a) *In malic acid solution, $[H^+] \approx [HA^-]$. This occurs in the graph at pH ≈ 3.33 (or log $[H^+]$ = -3.33), and $[HA^-] = 10^{-3.33}$ M= 4.5×10^{-4} M. At this pH, $[A^{2-}] = 10^{-5.00}$ M = 1.0×10^{-5} M, and $[H_2A] = 10^{-3.25}$ M = 5.6×10^{-4} M. (Since $[A^{2-}]$ is << $[HA^-]$, we can more accurately calculate that the relative $[H_2A] \approx C_{H2A} - [HA^-] = 1.0\times10^{-3} - 4.5\times10^{-4} - 5.5\times10^{-4}$ M, close to the graphical value.)*

(b) *In a malate solution, $[OH^-] \approx [HA^-]$, from the hydrolysis of A^{2-}. This occurs at pH = 8.05 (or log $[OH^-]$ = -5.95). At this pH, A^{2-} predominates (the hydrolysis is small), and $[HA^-] = 10^{-5.95}$ M – 1.1×10^{-6} M, $[H_2A] = 10^{-7.7}$ M = 2.0×10^{-8} M.*

66. $K_{a1} = [H^+][HA^-]/[H_2A]$ $\qquad K_{a2} = [H^+][A^{2-}]/[HA^-]$

In very acid solution, $[H_2A] \approx C_{H_2A} = 10^{-3}$ M

From K_{a1},

$[HA^-] = K_{a1}C_{HA}/[H^+]$

$\log [HA^-] = \log K_{a1}C_{H_2A} + pH = \log (4.0 \times 10^{-4})(10^{-3}) + pH = -6.40 + pH,$

slope = +1 for $pH < pK_{a1}$

At pH 1.4, then, $\log [HA^-] = -5$. *Check the figure.*

In very alkaline solution, $[A^{2-}] \approx C_{A^{2-}} = 10^{-3}$ M

From K_{a2},

$[HA^-] = C_{H_2A}[H^+]/K_{a2}$

$\log [HA^-] = \log (C_{H_2A}/K_{a2}) - pH = \log (10^{-3}/8.9 \times 10^{-6}) - pH = 2.05 - pH,$

slope = -1 for $pH > pK_{a2}$.

At pH 10.05, then, $\log [HA^-] = -8$. *Check the figure.*

67\. (a) At pH > pK_{a1}, assume $[H_2PO_4^{2-}] \approx C_{H_3PO_4} = 10^{-3}$ M (strictly speaking, this is true half-way between the two pK_a values, i.e., at the first end point of the H_3PO_4 titration).

From K_{a1}, we derive

$\log [H_3PO_4] = \log [H_2PO_4^{2-}]]/K_{a1} - pH$

$\approx \log (10^{-3})/(10^{-1.96}) - pH$

$\log [H_3PO_4] = -1.04 - pH$

At pH 2.96, log C = -4.00. This agrees with the H_3PO_4 curve.

(b) At pH between pK_{a2} and pK_{a3}, assume $[HPO_4^{2-}] \approx C_{H_3PO_4} = 10^{-3}$ M (true only at second end point of titration of H_3PO_4, midway between pK_{a2} and pK_{a3}).

From K_{a2}, we derive

$\log [HPO_4^{-}] = \log [HPO_4^{2-}]/K_{a2} - pH$

$\approx \log (10^{-3})/(10^{-7.12}) - pH$

$\log [H_2PO_4^{-}] = 4.12 - pH$

At pH 8.12, log C = -4.00, which agrees with the $H_2PO_4^{-}$ curve.

(c) At pH between pK_{a1} and pK_{a2}, assume $[H_2PO_4^{-}] \approx C_{H_3PO_4} = 10^{-3}$ M.

From K_{a2}, we derive

$\log [HPO_4^{2-}] = \log K_{a2}[HPO_4^{-}] + pH$

$\approx \log (10^{-7.12})(10^{-3}) + pH$

$\log [HPO_4^{2-}] = -10.12 + pH$

At pH 4.12, log C = -6.00, which agrees with the figure.

(d) At pH between pK_{a2} and pK_{a3}, assume $[HPO_4^{2-}] \approx C_{H_3PO_4} = 10^{-3}$, as in (b).

From K_{a3}, we derive

$\log [PO_4^{-3}] = \log K_{a3} [HPO_4^{2-}] + pH = \log (10^{-12.32})(10^{-3}) + pH$

$\log [PO_4^{3-}] = -15.32 + pH$

At pH 10.32, log C = -5.00, which agrees with the curve. Note that these expressions are approximations that hold closest to the midpoints between the pK_a values. The plots begin to curve at the extreme pH values.

68. *See the CD for the spreadsheet setup and chart of the H_3PO_4 log-log diagrams using alpha values.*

1. pH = 2. This is the pH range required to change the ratio of $[HIn]/[In^-]$ from 1/10 to 10/1, the ratios at which the eye perceives the color of only one form of the indicator.

2. Its transition range must fall within the sharp equivalence point break of the titration curve, i.e., pK_a (In) ≈ pH equivalence point.

3. At pH = pK_a.

4. Alkaline. Because at the end point, we have a solution of a salt of a weak acid, which is a Bronsted base.

5. For NH_3 vs HCl, methyl red. For HOAc vs NaOH, phenolphthalein.

6. CO_2 is boiled out, removing the HCO_3^-/CO_2 buffer system and allowing pH to increase to that of a HCO_3^- solution ($[H^+] = \sqrt{K_1K_2}$). The pH then drops sharply at the end point.

7. pK = 7-8

8. The difference in their K_a's must be > 10^4.

9. A primary standard is one whose purity is known. A secondary standard is one whose purity is unknown and is standardized against a primary standard.

10. A zwitterion forms in solution when an amphoteric substance, such as an amino acid, undergoes proton transfer from the acid group to the basic group.

11. Proteins typically contain 16% nitrogen.

12. Hydrochloric acid is the preferred titrant because most chlorides are soluble, and it has few side reactions.

13. $\underline{M}_{HCl}$ = [454.1 mg THAM/121.14 (mg/mmol)]/(35.37 mL) = 0.1060 mmol/mL

14. Reaction: $CO_3^{2-} + 2H^+ \rightarrow H_2CO_3$ ∴ mmol HCl = 2 x mmol Na_2CO_3

$\underline{M}_{HCl}$ = [(232.9 mg Na_2CO_3/105.99 mg/mmol) x 2 mmol HCl/mmol Na_2CO_3]/42.87 mL

$= 0.1025$ mmol/mL

15. $\underline{M}_{NaOH}$ = [859.2 mg KHP/204.23 mg/mmol]/32.67 mL = 0.1288 mmol/mL

16. 1 mmol HCl gives 1 mmol AgCl.

$\underline{M}_{HCl}$ = [168.2 mg/143.32 mg/mmol]/10.00 mL = 0.1174 mmol/mL

17. $B + H_2O = BH^+ + OH^-$

$pOH = pK_b + \log ([BH^+]/[B])$

To see acid form:

$pH = (14 - pK_b + \log (10/1) = 15 - pK_b$

To see base form:

$pH = (14 - pK_b) + \log (1/10) = 13 - pK_b$

$\Delta pH = (15 - pK_b) - (13 - pK_b) = 2$

The transition is around $pH = 14 - pK_b$.

18. 0 mL: pOH = -log 0.100 = 1.00; pH = 13.00

10.0 mL: mmol OH^- left = 0.100 $\underline{M}$ x 50.0 mL - 0.200 $\underline{M}$ x 10.0 mL = 3.00 mmol

$[OH^-]$ = 3.00 mmol/60.0 mL = 0.0500 $\underline{M}$

pOH = -log 0.0500 = 1.30; pH = 12.70

25.0 mL: All the OH^- has been converted to H_2O. pH = 7.00

30.0 mL: mmol excess H^+ = 5.0 mL x 0.200 $\underline{M}$ = 1.0 mmol/80 mL = 0.012_5 $\underline{M}$

pH = -log 0.012_5 = 1.90

19. 0 mL: $[H^+] = \sqrt{K_a[HA]} = \sqrt{2.0 \times 10^{-5} \times 0.200} = 2.0 \times 10^{-3}$ $\underline{M}$

pH = 2.70

10.0 mL: mmol A^- produced = 0.100 $\underline{M}$ x 10.0 mL = 1.00 mmol

mmol HA left = 0.200 $\underline{M}$ x 25.0 mL - 1.00 mmol = 4.00 mmol

$pH = -\log 2.0 \times 10^{-5} + \log (1.00/4.00) = 4.10$

25.0 mL: mmol A^- formed = 0.100 M x 25.0 mL = 2.50 mmol

mmol HA left = 5.00 mmol - 2.50 mmol = 2.50 mmol

$pH = 4.70 + \log (2.50/2.50) = 4.70$

50.0 mL: mmol A^- formed = 0.100 M x 50.0 mL = 5.00 mmol in 75.0 mL

All the HA has been converted to A^-. $[A^-] = 0.0667$ M

$[OH^-] = \sqrt{[(1.0 \times 10^{-14})/(2.0 \times 10^{-5})] \times 0.0667} = 5.77 \times 10^{-6}$ M

pOH = 5.24; pH = 8.76

60.0 mL: mmol excess OH^- = 0.100 M x 10.0 mL = 1.00 mmol in 85.0 mL

$[OH^-] = 0.0118$ M

pOH = 1.93; pH = 12.07

20. 0 mL: $[OH^-] = \sqrt{K_b[NH_3]} = \sqrt{1.75 \times 10^{-5} \times 0.100} = 1.32 \times 10^{-3}$ M

pOH = 2.88; pH = 11.12

10.0 mL: mmol NH_4^+ formed = 0.100 M x 10.0 mL = 1.00 mmol

mmol NH_3 left = 0.100 M x 50.0 mL - 1.00 mmol = 4.00 mmol

$pH = 14.00 - [-\log 1.75 \times 10^{-5}) + \log (1.00/4.00)]$ 9.84

25.0 mL: mmol NH_4^+ formed = 0.100 M x 25.0 mL = 2.50 mmol

mmol NH_3 left = 5.00 - 2.50 = 2.50 mmol

$pOH = 4.76 + \log (2.50/2.50) = 4.76$; pH = 9.24

50.0 mL: mmol NH_4^+ formed = 0.100 M x 50.0 mL = 5.00 mmol in 100 mL. All the NH_3 has been converted to NH_4^+. $[NH_4^+] = 0.0500$ M

$[H^+] = \sqrt{[(1.0 \times 10^{-14})/(1.75 \times 10^{-5})] \times 0.0500} = 5.35 \times 10^{-6}$ M

pH = 5.27

60.0 mL: mmol excess H^+ = 0.100 $\underline{M}$ x 10.0 mL = 1.00 mmol in 110 mL

$[H^+] = 0.00909$ $\underline{M}$; pH = 2.04

21. 0%: Calculate pH from ionization of first proton of H_2A. Since [HA] is just equal to 100 x K_{a_1}, the quadratic equation need not be used.

$[H^+] = \sqrt{K_{a_1}[H_2A]} = \sqrt{1.0 \times 10^{-3} \times 0.100} = 1.0 \times 10^{-2}$ $\underline{M}$

pH = 2.00

25.0%: This is halfway to the first equivalence point and $[H_2A] = [HA^-]$
(50 mL)

(0.100 $\underline{M}$ x 50.0 mL = 5.00 mmol each)

$pH = pK_{a_1} + \log([HA^-]/[H_2A]) = 3.00 + \log(5.00/5.00) = 3.00$

50.0%: This is at the 1st equivalence point. All H_2A has been converted to
(100 mL)

HA^- (= 10.0 mmol in 200 mL).

$[H^+] = \sqrt{K_{a_1}K_{a_2}} = \sqrt{1.0 \times 10^{-3} \times 1.0 \times 10^{-7}} = 1.0 \times 10^{-5}$ $\underline{M}$

pH = 5.00

75.0%: This is halfway between the two equivalence points, and $[HA^-] = [A^{2-}]$
(150 mL)

(5.00 mmol each)

$pH = pK_{a_2} + \log([A^{2-}]/[HA^-]) = 7.00 + \log(5.00/5.00) = 7.00$

100.0%: All the HA^- has been converted to A^{2-} (= 10.0 mmol in 300 mL)
(200 mL)

$[A^2] = 0.0333$ $\underline{M}$

$A^{2-} + H_2O = HA^- + OH^-$

$[HA^-][OH^-]/[A^{2-}] = (K_w/K_{a_2}) = (1.0 \times 10^{-14}/1.0 \times 10^{-7}) = 1.0 \times 10^{-7}$

$[OH^-] = \sqrt{1.0 \times 10^{-7} \times [A^{2-}]} = \sqrt{1.0 \times 10^{-7} \times 0.0333} = 5.8 \times 10^{-5}$ $\underline{M}$

pOH = 4.24; pH = 9.76

125.0%: mmol excess OH^- = 0.100 $\underline{M}$ x 50 mL = 5.0 mmol in 350 mL (250 mL)

(The hydrolysis of A^{2-} is negligible in the presence of added OH^-.)

$[OH^-] = 0.0143\ \underline{M}$

pOH = 1.84; pH = 12.16

22. 0%: $[H^+] = \sqrt{K_{a_2}K_{a_3}} = \sqrt{(7.5 \times 10^{-8})(4.8 \times 10^{-13})} = 1.9_0 \times 10^{-10}\ \underline{M}$

pH = 9.72

25.0%: mmol $H_2PO_4^-$ formed = 0.100 $\underline{M}$ x 25.0 mL = 2.50 mmol

mmol HPO_4^{2-} left = 0.100 $\underline{M}$ x 100 mL - 2.50 mmol = 7.50 mmol

$pH = pK_{a_2} + \log [HPO_4^{2-}]/[H_2PO_4^-] = 7.12 + \log (7.50/2.50) = 7.60$

50.0%: mmol $H_2PO_4^-$ formed = 0.100 $\underline{M}$ x 50.0 mL = 5.00 mmol

mmol HPO_4^{2-} left = 10.0 mmol - 5.00 mmol = 5.0 mmol

$pH = pK_{a_2} = 7.12$

75%: mmol $H_2PO_4^-$ formed = 0.100 $\underline{M}$ x 75.0 mL = 7.50 mmol

mmol HPO_4^{2-} left = 10.0 - 7.50 = 2.5 mmol

$pH = 7.12 + \log (2.5/7.5) = 6.64$

100%: mmol $H_2PO_4^-$ formed = 0.100 $\underline{M}$ x 100 mL = 10.0 mmol in 200 mL

All the HPO_4^{2-} has been converted to $H_2PO_4^-$. $[H_2PO_4^-] = 0.0500\ \underline{M}$

$[H^+] = \sqrt{K_{a_1}K_{a_2}} = \sqrt{(1.1 \times 10^{-2})(7.5 \times 10^{-8})} = 2.8 \times 10^{-5}\ \underline{M}$

pH = 4.54

150%: mmol H_3PO_4 formed = 50.0 mL x 0.100 $\underline{M}$ = 5.00 mmol

mmol $H_2PO_4^-$ left = 10.0 - 5.00 = 5.0 mmol

$pH = pK_{a_1} + \log [H_2PO_4^-]/[H_3PO_4] = 1.96 + \log (5.0/5.0) = 1.96$

23. % KH_2PO_4 = {[25.6 mL x 0.112 (mmol/mL) x 136 (mg/mmol)]/(492 mg)} x 100%

= 79.2%

24. 90.0% = {[0.155 (mmol/mL) x x mL x 2 (mmol LiOH/mmol H_2SO_4)

x 23.95 (mg/mmol)]/(293 mg)} x 100%

x = 35.5 mL

25. [0.250 (mmol/mL) x 25.0 mL - 0.250 (mmol/mL) x 9.26 mL] x 56.1 mg/mmol

= 221 mg KOH reacted. Saponification no. = 221 mg KOH/1.10 g fat = 201 mg/g

{[3.94 mmol KOH reacted x 1/3 (mmol fat/mmol KOH)

x $f.w._{fat}$]/(1100 mg)} x 100% = 100% fat

$f.w._{fat}$ = 838 mg/mmol

26. mmol alanine = mmol N = 2 mmol H_2SO_4 reacted

∴ % alanine = {[(mmol H_2SO_4 - mmol NaOH x 1/2) x 2 x $CH_3CH(NH_2)COOH$]

/(mg sample)} x 100% = {[(0.150 x 50.0 - 0.100 x 9.0 x 1/2) x 2 x 89.10]

/(2000)} x 100% = 62.8%

27. Standardization: net reaction is

$(NH_4)_2SO_4 + 2HCl \quad 2NH_4Cl + H_2SO_4$

∴ mmol HCl = 2 x mmol $(NH_4)_2SO_4$

$\underline{M}_{HCl}$ x 33.3 mL = 2 x [(330 mg)/(132.1 mg/mmol)]

$\underline{M}_{HCl}$ = 0.150 $\underline{M}$

g protein/100 mL =

[(0.150 $\underline{M}$ x 15.0 mL x 0.01401 g N/mmol x 6.25 g protein/g N)/(2.00 mL)] x 100 mL

= 9.85 g %

28. mmol H_3PO_4 = 0.200 $\underline{M}$ x 10.0 mL = 2.00 mmol

mmol HCl = 0.200 $\underline{M}$ (25.0 - 10.0 mL) = 3.00 mmol

$\therefore$ H_3PO_4 = 2.00 mmol/100 mL = 0.0200 $\underline{M}$

HCl = 3.00 mmol/100 mL = 0.0300 $\underline{M}$

29. The first end point corresponds to adding 1 H^+ to CO_3^{2-} and the second end point (28.1 mL later) corresponds to adding 1 more H^+ to CO_3^{2-} plus titration of the original HCO_3^-.

mmol CO_3^{2-} = 0.109 $\underline{M}$ x 15.7 mL = 1.71 mmol

mmol HCO_3^- = 0.109 $\underline{M}$ (28.1 - 15.7) mL = 1.35 mmol

% Na_2CO_3 = [(1.71 mmol x 106.0 mg/mmol)/(527 mg)] x 100% = 34.4%

% $NaHCO_3$ = [(1.35 mmol x 84.0 mg/mmol)/(527 mg)] x 100% = 21.5%

30. mg Na_2CO_3 = 0.250 $\underline{M}$ x 15.2 mL x 106 mg/mmol = 403 mg

mg NaOH = 0.250 $\underline{M}$ x (26.2 - 15.2) mL x 40.0 mg/mmol = 110 mg

31. $\underline{M}_{HCl}$ = [(477 mg Na_2CO_3)/(106.0 mg/mmol) x 2 (mmol HCl/mmol Na_2CO_3)]/(30.0 mL) = 0.300 mmol/mL

The first end point is shorter than the second and corresponds to titration of CO_3^{2-} only to HCO_3^-. The second end point corresponds to titration of CO_3^{2-} (in the form of HCO_3^-) plus HCO_3^- originally present to H_2CO_3.

mmol Na_2CO_3 = 15.0 mL x 0.300 $\underline{M}$ x 1(mmol Na_2CO_3/mmol HCl) = 4.50 mmol

The same volume is used to titrate the CO_3^{2-} to the second end point:

mmol $NaHCO_3$ = (35.0 - 15.0) mL x 0.300 $\underline{M}$ x 1(mmol $NaHCO_3$/mmol HCl) = 6.00 mmol

32. In this case, the first end point is longer than the second and corresponds to titration of CO_3^{2-} to HCO_3^- plus OH^- to H_2O. The second end point corresponds to titration of only CO_3^{2-} (in the form of HCO_3^-) to H_2CO_3.

mmol Na_2CO_3 = 10.0 mL x 0.300 M x 1(mmol Na_2CO_3/mmol HCl) = **3.00 mmol**

mmol NaOH = (15.0 - 10.0) mL x 0.300 M x 1(mmol NaOH/mmol HCl) = 1.5_0 mmol

33. Let x = mg $BaCO_3$. Then mg Li_2CO_3 = 150 - x

$mmol_{HCl} = mmol_{BaCO_3} \times 2 + mmol_{Li_2CO_3} \times 2$

$= [mg_{BaCO_3}/f.w._{BaCO_3}\ (mg/mmol)] \times 2\ (mmol\ HCl/mmol\ BaCO_3)$

$+ [mg_{Li_2CO_3}/f.w._{Li_2CO_3}\ (mg/mmol)] \times 2\ (mmol\ HCl/mmol\ Li_2CO_3)$

$0.120\ M\ (mmol/mL) \times 25.0\ mL = (x\ mg_{BaCO_3}/197.35\ mg/mmol) \times 2$

$+ [(150 - x)\ mg_{Li_2CO_3}/73.89\ mg/mmol] \times 2$

x = 62.6 mg $BaCO_3$

% $BaCO_3$ = (62.6)/(150 mg) x 100% = 41.7%

34. Let x = mg H_3PO_4. Then mg P_2O_5 = 405 - x

$mmol_{NaOH} = mmol_{H_3PO_4} \times 2\ (mmol\ NaOH/mmol\ H_3PO_4) + mmol\ P_2O_5$

$\times 4\ (mmol\ NaOH/mmol\ P_2O_5) = [mg_{H_3PO_4}/f.w._{H_3PO_4}\ (mg/mmol)] \times 2$

$+ [mg_{P_2O_5}/f.w._{P_2O_5} (mg/mmol)] \times 4$

$0.250\ M\ (mmol/mL) \times 42.5\ mL = [x\ mg_{H_3PO_4}/98.0\ (mg/mmol)] \times 2$

$+ [(405 - x)\ mg_{P_2O_5}/141.9\ (mg/mmol)] \times 4$

x = 10_2 mg H_3PO_4

% H_3PO_4 = (10_2 mg/405 mg) x 100% = $25._2$

35. See CD for spreadsheet and graph of titration curve.

1. *A chelating agent is a type of complexing agent in which the complexing molecule has two or more complexing groups.*

2. *Chelation titration indicators are themselves chelating agents that form a weaker chelate with the titrate metal ion than does the titrant. The titrant displaces all the indicator from the metal ion at the equivalent point, causing the color to revert to that of the uncomplexed indicator.*

3. *The indicator forms too weak a chelate with calcium to give a sharp end point. The added magnesium combines with the indicator, and the end point occurs when the EDTA displaces the last of the indicator from the magnesium. The EDTA is standardized after adding the magnesium in order to correct for the decrease in the effective molarity of EDTA due to reaction with magnesium.*

4. *After mixing, we start with 0.0050 M* Ca^{2+} *and 1.0 M* NO_3^-. *We neglect the amount of reacted* NO_3^-. *At equilibrium we have:*

$$Ca^{2+} + NO_3^- = Ca(NO_3)^+$$

$$x \qquad 1.0 + x \qquad 0.0050 - x$$

Try neglecting x compared to 1.0, but not compared to 0.0050.

$$[Ca(NO_3)^+]/([Ca^{2+}][NO_3^-]) = K_f = 2.0$$

$(0.0050-x)/[(x)(1.0)] = 2.0$

$x = [Ca^{2+}] = 0.0017$ M

$[Ca(NO_3)^+] = 0.0050 - 0.0017 = 0.0033$ M

5. Let en = $NH_2CH_2CH_2NH_2$

en + Ag^+ = $Ag(en)^+$

x x 0.10 - x

$[Ag(en)^+]/[en][Ag^+] = 5.0 \times 10^4$

Assume x is negligible compared to 0.10. Then,

$(0.10)/(x)(x) = 5.0 \times 10^4$

$x = 1.4 \times 10^{-3}$ M $= [Ag^+]$

6. en + Ag^+ = $Ag(en)^+$

0.10 + x x 0.10 - x

Again, x << 0.10, so

$(0.10)/[(0.10)(x)] = 5.0 \times 10^4$

$x = 2.0 \times 10^{-5}$ M $= [Ag^+]$

7. The equilibria are

$Ag(S_2O_3)_2^{3-} = Ag(S_2O_3)^- + S_2O_3^{2-}$

$Ag(S_2O_3)^- = Ag^+ + S_2O_3^{2-}$

For a complex with two ligands [similar to $Ag(NH_3)_2^+$ -- see Equations 9.20 - 9.22], we calculate that

$\beta_0 = 1/(K_{f_1}K_{f_2}[S_2O_3^{2-}]^2 + K_{f_1}[S_2O_3^{2-}] + 1) = 3$

(for 1.00 M $S_2O_3^{2-}$)

$\beta_1 = (K_{f_1}[S_2O_3^{2-}])/(K_{f_1}K_{f_2}[S_2O_3^{2-}]^2 + K_{f_1}[S_2O_3^{2-}] + 1) = 2.2_8 \times 10^{-15}$

$\beta_2 = (K_{f_1}K_{f_2}[S_2O_3^{2-}]^2)/(K_{f_1}K_{f_2}[S_2O_3^{2-}]^2 + K_{f_1}[S_2O_3^{2-}] + 1) = 1.00$

$[Ag^+] = C_{Ag}\beta_0 = (0.0100)(3.4_5 \times 10^{-14}) = 3.4_5 \times 10^{-16}$ $\underline{M}$

$[Ag(S_2O_3)^-] = C_{Ag}\beta_1 = (0.0100)(2.2_8 \times 10^{-5}) = 2.2_8 \times 10^{-7}$ $\underline{M}$

$[Ag(S_2O_3)_2^{3-}] = C_{Ag}\beta_2 = (0.0100)(1.00) = 1.00 \times 10^{-2}$ $\underline{M}$

The silver exists almost exclusively as the di-thiosulfate complex.

8. (a) From Equation 9.12 ,

$1/\alpha_4 = 1 + (10^{-3})/(5.5 \times 10^{-11}) + [(10^{-3})^2]/[(6.9 \times 10^{-7})(5.5 \times 10^{-11})]$

$+ [(10^{-3})^3]/[(2.2 \times 10^{-3})(6.9 \times 10^{-7})(5.5 \times 10^{-11})]$

$+ [(10^{-3})^4]/[(1.0 \times 10^{-2})(2.2 \times 10^{-3})(6.9 \times 10^{-7})(5.5 \times 10^{-11})]$

$\alpha_4 = 2.5_3 \times 10^{-11}$

$K_f' = K_f\alpha_4 = 1.10 \times 10^{18} \times 2.5_3 \times 10^{-11} = 2.7_8 \times 10^7$

(b) $1/\alpha_4 =$

$1 + (10^{-10})/(5.5 \times 10^{-11}) + [(10^{-10})^2]/[(6.9 \times 10^{-7})(5.5 \times 10^{-11})]$

$+ [(10^{-10})^3]/[(2.2 \times 10^{-3})(6.9 \times 10^{-7})(5.5 \times 10^{-11})]$

$+ [(10^{-10})^4]/[(1.0 \times 10^{-2})(2.2 \times 10^{-3})(6.9 \times 10^{-7})(5.5 \times 10^{-11})]$

$\alpha_4 = 0.35_5$

$K_f' = K_f\alpha_4 = 1.10 \times 10^{18} \times 0.35_5 = 3.9_0 \times 10^{17}$

9. (a) $K_f' = 2.7_8 \times 10^7$

(1) $pPb = -\log [Pb^{2+}] = -\log 2.50 \times 10^{-2} = 1.60$

(2) mmol Pb^{2+} start = 0.0250 $\underline{M}$ x 50.0 mL = 1.25 mmol

mmol EDTA added = 0.0100 $\underline{M}$ x 50.0 mL = 0.50 mmol

= mmol PbY^{2-} formed ---------------------

mmol Pb^{2+} left = 0.75 mmol

Neglecting dissociation of PbY^{2-},

$pPb = -\log (0.75 \text{ mmol}/100 \text{ mL}) = 2.12$

(3) Stoichiometric amounts of Pb^{2+} and EDTA =

$1.25 \text{ mmol } PbY^{2-}/175 \text{ mL} = 7.14 \times 10^{-3}$ $\underline{M}$

$[PbY^{2-}]/[Pb^{2+}]C_{H_4Y} = 2.7_8 \times 10^7$;

$(7.14 \times 10^{-3} - x)/[(x)(x)] = 2.7_8 \times 10^7 \approx (7.14 \times 10^{-3})/(x^2)$

$x = 1.60 \times 10^{-5}$ $\underline{M}$ $= [Pb^{2+}]$; $pPb = -\log 1.60 \times 10^{-5} = 4.80$

(4) $[PbY^{2-}] = 1.25 \text{ mmol}/250 \text{ mL} = 5.00 \times 10^{-3}$ $\underline{M}$

mmol excess EDTA = 0.0100 $\underline{M}$ x 200 mL - 0.0100 $\underline{M}$ x 125 mL = 0.75 mmol

$C_{H_4Y} = 0.75 \text{ mmol}/250 \text{ mL} = 3.0_0 \times 10^{-3}$ $\underline{M}$

$(5.00 \times 10^{-3})/[Pb^{2+}](3.0_0 \times 10^{-3}) = 2.7_8 \times 10^7$;

$[Pb^{2+}] = 6.0_0 \times 10^{-8}$ $\underline{M}$

$pPb = -\log 6.0_0 \times 10^{-8} = 7.22$

(b) $3.9_0 \times 10^{17}$

(1) pPb = 1.60 [as in (a)]

(2) pPb = 2.12 [as in (a)]

(3) From (a): $[PbY^{2-}] = 7.14 \times 10^{-3}$ $\underline{M}$

$(7.14 \times 10^{-3})/(x^2) \approx 3.9_0 \times 10^{17}$; $x = 1.3_5 \times 10^{-10} = [Pb^{2+}]$

$pPb = -\log 1.3_5 \times 10^{-10} = 9.87$

(4) From (a): $C_{H_4Y} = 3.0_0 \times 10^{-3}$ $\underline{M}$, $[PbY^{2-}] = 5.00 \times 10^{-3}$ $\underline{M}$

$(5.00 \times 10^{-3})/[Pb^{2+}](3.0_0 \times 10^{-3}) = 3.9_0 \times 10^{17}$;

$[Pb^{2+}] = 4.2_7 \times 10^{-18}$ $\underline{M}$

$pPb = -\log 4.2_7 \times 10^{-18} = 17.37$

10. From Problem *8*, $\alpha_4 = 2.5_3 \times 10^{-11}$

$\therefore K_f' = K_f\alpha_4 = 5.0 \times 10^{10} \times 2.5_3 \times 10^{-11} = 1.2_6$

Hence, at pH 3, the Ca - EDTA chelate is very weak. In view of the large differences in the K_f' values for calcium and lead, it should be possible to titrate lead in the presence of calcium.

11. 0.05000 M x 500.0 mL = (x mg $Na_2H_2Y.2H_2O$)/(372.23 mg/mmol)

x = 9306 mg = 9.306 g $Na_2H_2Y.2H_2O$

12. M_{EDTA} x 38.26 mL = (398.2 mg)/(100.09 mg/mmol)

$M_{EDTA} = 0.1039_8$

13. Titer = mg $CaCO_3$/mL EDTA. Since the reaction is 1:1, mmol $CaCO_3$ = mmol EDTA

mmol $CaCO_3 = M_{EDTA} \times mL_{EDTA}$

(mg $CaCO_3$)/[100.1 (mg/mmol)] = 0.1000 M x 1.000 mL = 0.1000 mmol

mg $CaCO_3$ = 0.1000 x 100.1 = 10.01 mg/mL EDTA

14. Water hardness = mg $CaCO_3$/L H_2O. The milligrams $CaCO_3$ in 100.0 mL (1 dL) = 1/10th the water hardness. From Problem *13*, 0.100 M EDTA is equivalent to 10.01 mg $CaCO_3$/mL, so 0.01000 M EDTA is equivalent to 1.001 mg $CaCO_3$/mL.

mL_{EDTA} x 1.001 (mg $CaCO_3$/mL) x 1) dL/L = mg $CaCO_3$/L H_2O = water hardness

Water hardness/mL EDTA = 10.01 (mg $CaCO_3$/L H_2O)/mL EDTA

15. M_{Zn} = [632 mg/65.4 (mg/mmol)]/(1000 mL) = 0.00966 mmol/mL

M_{EDTA} = [10.0 mL_{Zn} x 0.00966 (mmol/mL)]/(10.8 mL_{EDTA}) = 0.00894 mmol/mL

mg_{Ca} = 0.00894 mmol/mL EDTA x 12.1 mL EDTA x 40.1 Ca/mmol = 4.34 mg

ppm_{Ca} = mg/kg = 4.34 mg/1.50 x 10^{-3} kg = 2.89 x 10^3 ppm

16. $[0.00122\ \underline{M} \times 0.203\ mL \times 40.1\ mg/mmol]/10^{-3}\ (dL/100\ \mu L) = 9.93\ mg\ Ca/dL$ serum

$9.93\ mg/dL = 99.3\ mg/L$

$(99.3\ mg/L)/(40.1/2\ mg/meq) = 4.95\ meq/L$

17. % KCN = $\{[0.1025\ \underline{M} \times 34.95\ mL \times 2\ (mmol\ CN^-/mmol\ Ag^+)$

$\times 65.12\ mg/mmol]/(472.3\ mg)\} \times 100\% = 98.79\%$

18. EDTA titer = $2.69\ mg\ CaCO_3/mL$

$\underline{M}_{EDTA} = [2.69\ mg\ CaCO_3/100.1\ (mg\ CaCO_3/mmol)]/(1\ mL\ EDTA) = 0.0269\ mmol/mL$

mg_{Cu} titrated = $0.0269\ mmol/mL\ EDTA \times 2.67\ mL\ EDTA \times 63.5\ mg\ Cu/mmol$

$= 4.56\ mg$

$ppm_{Cu} = mg/L = [4.56\ (mg/50\ mL) \times 2]/(3\ L) = 3.04\ mg/L$

19. $\underline{M}_{Hg(NO_3)_2} \times 1.12\ mL = 0.0108\ \underline{M} \times 2.00\ mL \times 1/2\ (mmol\ Hg(NO_3)_2/mmol\ NaCl)$

$\underline{M}_{Hg(NO_3)_2} = 0.00964\ \underline{M}$

One-half mL serum was diluted to 5.00 mL and 2.00 mL (40.0%) of this was taken for titration, or 0.200 mL serum.

$mg\ Cl^-/L = [0.00964\ \underline{M}\ 1.23\ mL \times 2\ mmol\ Cl^-/mmol\ Hg(NO_3)_2$

$\times 1\ mg/mmol \times 10^3\ mL/L]/(0.200\ mL\ serum) = 119\ mg/L$ serum

20. See CD for spreadsheet and graph of $\log K_f'$ values.

21. See CD for spreadsheet and graph of titration curve.

22. See CD for spreadsheet and graph of beta-values.

1. *The solution is adjusted for optimum precipitating conditions, and the analyte is precipitated, digested to obtain a pure and filterable precipitate, filtered, washed to remove impurities (with a volatile electrolyte to prevent peptization), dried or ignited to a weighable form, and weighed in order to calculate the quantity of analyte.*
2. *The relative supersaturation (Q-S)/(S). Q = concentration of mixed reagents before precipitation, S = solubility of precipitate.*
3. *Most favorable precipitates are obtained with minimal supersaturation, and the von Weimarn ratio predicts that favorable precipitation conditions are obtained by precipitating from dilute solution, slowly, with stirring (low Q), and from hot solution (high S).*
4. *The process of allowing a precipitate to stand in the presence of the mother liquor, often at elevated temperature, in order to obtain purer and larger crystals. Surface impurities and small crystals dissolve and the latter reprecipitate on the larger crystals, resulting in more perfect and larger crystals.*
5. *See answer to Question 3, and precipitate at as low a pH as possible to maintain quantitative precipitation.*
6. *Coprecipitation is the carrying down with the precipitate of normally soluble constituents in the solution. In occlusion, foreign ions are trapped in the crystal as it grows. These are difficult to get rid of, and reprecipitation may be required. Impurities adsorbed on the surface of the precipitate can often be removed by digestion and/or washing. Post precipitation is the slow precipitation (after a period of time) of a normally insoluble precipitate. It can be minimized by filtering as soon as possible. Isomorphous replacement is the formation of mixed crystals of two salts that are chemically similar. It is difficult to eliminate.*

7. *In order to remove the mother liquor and nonvolatile surface impurities and replace them by a volatile electrolyte.*

8. *In order to prevent peptization (formation of colloidal particles) of the precipitate. The electrolyte must be volatile at the temperature of drying or ignition and must not dissolve the precipitate.*

9. *Organic precipitation agents produce precipitates with very low solubility in water that have very favorable gravimetric factors since the organic reagents have high molecular weights. Depending on the reagent, selectivity can be high. Solubility can be adjusted by pH control.*

10. g Na = g Na_2SO_4 x ($2Na/Na_2SO_4$) = 50.0 x [2(22.99)/142.0] = 16.2 g

11. g $BaSO_4$ = g Na_2SO_4 x ($BaSO_4/Na_2SO_4$) = 50.0 x (233.4/142.0) = 82.2 g

12. (As_2O_3)/($2Ag_3AsO_4$) = (197.8)/2(462.5) = 0.2138

($2FeSO_4$)/(Fe_2O_3) = [2(151.9)]/159.7 = 1.902

(K_2O)/[$2KB(C_6H_5)_4$] = (94.20)/2(358.3) = 0.1314

($3SiO_2$)/($KAlSi_3O_8$) = [3(60.08)]/278.4 = 0.6474

13. g CuO = 1.00 x [(4CuO)/$Cu_3(AsO_3)_2 \cdot 2As_2O_3 \cdot Cu(C_2H_3O_2)_2$]

= 1.00 x [4(79.54)]/1.013 = 0.314 g

g As_2O_3 = 1.00 x [($3As_2O_3$)/$Cu_3(AsO_3)_2 \cdot 2As_2O_3 \cdot Cu(C_2H_3O_2)_2$]

= 1.00 x [3(197.8)]/1,013 = 0.586 g

14. % KBr = ([mg AgBr x (KBr)/(AgBr)]/mg sample) x 100%

= ([814.5 x (119.01)/(187.78)]/523.1) x 100% = 98.68%

15. g Fe = 0.4823 x 0.9989 = 0.4818 g

g Fe_2O_3 = g Fe x (Fe_2O_3)/(2 Fe) = 0.4818 g x (159.69)/(2 x 55.847) = 0.6888 g

16. % Al = (g $Al(C_9H_6ON)_3$ x [$Al/Al(C_9H_6ON)_3$]/g sample) x 100%

$= [0.1862\ g(26.98/459.5)/1.021\ g] \times 100\% = 1.071\%$

17. Calculate the amount of iron required to give 100.0 mg Fe_2O_3.

$mg\ Fe = mg\ Fe_2O_3 \times (2\ Fe/Fe_2O_3) = 100.0\ mg \times (2 \times 55.85/159.7) = 69.94\ mg$

The minimum iron content is 11%, so

$69.94\ mg \times (100/11) = 636$ mg minimum sample needed.

18. $g\ MgCl_2 = 0.12\ g \times 0.95 = 0.11_4\ g$

$g\ Cl^- = 0.11_4 \times (2\ Cl/MgCl_2) = 0.11_4 \times (2 \times 35.4/95.2) = 0.085\ g$

$mmol\ Cl^- = mmol\ Ag^+ = (85\ mg)/(35\ mg/mmol) = 2.4\ mmol$

Total mmol Ag^+ to be added $= 2.4 + 0.2 = 2.6$ mmol

$0.100\ \underline{M} \times x\ mL = 2.6\ mmol$

$x = 26$ mL $AgNO_3$

19. $2NH_3 \rightarrow (NH_4)_2PtCl_6 \rightarrow Pt$

$\therefore \%\ NH_3 = [g\ Pt \times (2NH_3/Pt)/g\ sample] \times 100\%$

$= [0.100 \times (2 \times 17.03)/195.1)/1.00] \times 100\% = 1.75\%$

20. $\%\ Cl = [g\ AgCl \times (Cl/AgCl)/x\ g] \times 100\%$

Since % Cl = g AgCl, these cancel. Thus,

$x = (35.45/143.3) \times 100 = 24.74\ g$

21. Assume 1.000 g of $BaSO_4$, so the % FeS_2 is 10.00.

$[1.000 \times (FeS_2/2BaSO_4)/x\ g] \times 100\% = 10.00\%$

$[1.000 \times (120.0)/2(233.4)/x\ g] \times 100\% = 10.00\%$

$x = 2.571$ g of ore

22. $x = g\ BaO$ $\qquad$ $y = g\ CaO$

(1) $x + y = 2.00\ g$

(2) $x\ (BaSO_4/BaO) + y\ (CaSO_4/CaO) = 4.00\ g$

$x\ (233.4/153.3) + y\ (136.1/56.08) = 4.00\ g$

$1.52x + 2.43y = 4.00\ g$

Solution of the simultaneous equations gives:

$x = 0.95\ g\ BaO$ $\qquad$ $y = 1.05\ g\ CaO$

% Ba = [0.95 x (137.3/153.3)/2.00] x 100% = 42.5%

% Ca = [1.05 x (40.08/56.08)/2.00] x 100% = 37.5%

23. Assume 100.0 g sample. $x = g\ CaSO_4$

$g\ CaSO_4 + g\ BaSO_4 = 100.0$

$$x + \underbrace{\underbrace{x\ (Ca/CaSO_4)}_{g\ Ca}(1/2)}_{g\ Ba}(BaSO_4/Ba) = 100.0$$

$x + x\ (40.08/136.1)(1/2)(233.4/137.3) = 100.0$

$x = 79.98$

% $CaSO_4$ = (79.98/100.0) x 100% = 79.98%

24. There are two unknowns, and so there must be two equations.

$x = g\ AgCl$ $\qquad$ $y = g\ AgBr$

(1) $x + y = 2.00$

(2) g Ag from AgCl + g Ag from AgBr = 1.300

$x\ (Ag/AgCl) + y\ (Ag/AgBr) = 1.300\ g$

$x\ (107.9/143.3) + y\ (107.9/187.8) = 1.300\ g$

$0.7530x + 0.5745y = 1.300\ g$

Solution of the simultaneous equations gives:

$x = 0.846\ g\ AgCl$

$y = 1.154\ g\ AgBr$

25. (a) $K_{sp} = [Ag^+][SCN^-]$

(b) $K_{sp} = [La^{3+}][IO_3^-]^3$

(c) $K_{sp} = [Hg_2^{2+}][Br^-]^2$

(d) $K_{sp} = [Ag^+][Ag(CN)_2^-]$

(e) $K_{sp} = [Zn^{2+}]^2[Fe(CN)_6^{4-}]$

(f) $K_{sp} = [Bi^{3+}]^2[S^{2-}]^3$

26. $(7.76 \times 10^{-3}$ g/L)/(590 g/mol) $= 1.32 \times 10^{-5}$ $\underline{M}$

$BiI_3 = Bi^{3+} + 3I^-$

$[Bi^{3+}] = 1.32 \times 10^{-5}$ $\underline{M}$

$[I^-] = 3 \times 1.32 \times 10^{-5}$ $\underline{M}$ $= 3.96 \times 10^{-5}$ $\underline{M}$

$K_{sp} = [Bi^{3+}][I^-]^3 = (1.32 \times 10^{-5})(3.96 \times 10^{-5})^3 = 8.20 \times 10^{-19}$

27. $Ag_2CrO_4 = 2Ag^+ + CrO_4^{2-}$

$\qquad\qquad\quad 2s \qquad s$

$[Ag^+]^2[CrO_4^{2-}] = 1.1 \times 10^{-12}$

$(2s)^2(s) = 1.1 \times 10^{-12}$

$s = 6.5 \times 10^{-5}$

$[Ag^+] = 2 \times 6.5 \times 10^{-5} = 1.3 \times 10^{-4}$ $\underline{M}$

$[CrO_4^{2-}] = 6.5 \times 10^{-5}$ $\underline{M}$

28. mmol Ba^{2+} = 0.100 x 25.0 = 2.50 mmol

mmol CrO_4^{2-} = 0.200 x 15.0 = 3.00 mmol

excess CrO_4^{2-} = 3.00 - 2.50 = 0.50 mmol/40 mL = 0.012_5 $\underline{M}$

$BaCrO_4 = Ba^{2+} + CrO_4^{2-}$

s (s+0.0125)

$[Ba^{2+}][CrO_4^{2-}] = 2.4 \times 10^{-10}$

$[Ba^{2+}] = (2.4 \times 10^{-10})/(0.012_5) = 1.9 \times 10^{-8}$ $\underline{M}$

29. $[Ag^+]^3[PO_4^{3-}] = 1.3 \times 10^{-20}$

$[PO_4^{3-}] = (1.3 \times 10^{-20})/(0.10)^3 = 1.3 \times 10^{-17}$ $\underline{M}$

30. For PO_4^{3-}:

$[Ag^+]^3[PO_4^{3-}] = 1.3 \times 10^{-20}$

$[Ag^+] = \sqrt[3]{(1.3 \times 10^{-20})/(0.10)} = 5.1 \times 10^{-7}$ $\underline{M}$

For Cl^-:

$[Ag^+][Cl^-] = 1.0 \times 10^{-10}$

$[Ag^+] = (1.0 \times 10^{-10})/(0.10) = 1.0 \times 10^{-9}$ $\underline{M}$

31. $Al(OH)_3 \rightleftharpoons Al^{3+} + 3OH^-$

$[Al^{3+}][OH^-]^3 = 2 \times 10^{-32}$

$(0.10)[OH^-]^3 = 2 \times 10^{-32}$

$[OH^-] = \sqrt[3]{2 \times 10^{-32}/0.10 = 6 \times 10^{-11}\ \underline{M}}$

$pOH = -\log 6 \times 10^{-11} = 10.2$

$pH = 14.0 - 10.2 = 3.8$

Hence, $Al(OH)_3$ will form when the pH just exceeds 3.8.

32. $Ag_3AsO_4 = 3Ag^+ + AsO_4^{3-}$

$3s \quad s$

$[Ag^+]^3[AsO_4^{3-}] = 1.0 \times 10^{-22}$

$(3s)^3(s) = 1.0 \times 10^{-22}$

$s = 1.4 \times 10^{-6}$

$[Ag_3AsO_4] = 1.4 \times 10^{-6}$ $\underline{M}$

1.4×10^{-6} mol/L $\times$ 0.25 L $\times$ 463 g/mol = 1.6×10^{-4} g Ag_3AsO_4

33. Let s = solubility of Ag_2CrO_4

$Ag_2CrO_4 = 2Ag^+ + CrO_4^{2-}$

$2s \quad (s + 0.10)$

$[Ag^+]^2[CrO_4^{2-}] = 1.1 \times 10^{-12}$

$(2s)^2(0.10 + s) = 1.1 \times 10^{-12}$

Assume s is negligible compared to 0.10. Then,

$0.40\ s^2 = 1.1 \times 10^{-12}$

$s = 1.7 \times 10^{-6}$ $\underline{M}$

34. $AB = A^{2+} + B^{2+}$ | $AC_2 = A^{2+} + 2C^-$

$s \quad s$ | $s \quad 2s$

$[A^{2+}][B^{2+}] = 4 \times 10^{-18}$ | $[A^{2+}][C^-]^2 = 4 \times 10^{-18}$

$(s)(s) = 4 \times 10^{-18}$ | $(s)(2s)^2 = 4 \times 10^{-18}$

$s = 2 \times 10^{-9}$ $\underline{M}$ | $s = 1 \times 10^{-6}$ $\underline{M}$

$\therefore$ Compound AC_2 is more soluble.

35. $Bi_2S_3 = 2Bi^{3+} + 3S^{2-}$

Let s be the solubility. Then $[Bi^{3+}] = 2s$, $[S^{2-}] = 3s$.

$[Bi^{3+}]^2[S^{2-}]^3 = 1 \times 10^{-97}$

$(2s)^2(3s)^3 = 1 \times 10^{-97}$

$s = 1._6 \times 10^{-20}$ $\underline{M}$

$HgS = Hg^{2+} + S^{2-}$

Let s be the solubility. Then $[Hg^{2+}] = [S^{2-}] = s$.

$[Hg^{2+}][S^{2-}] = 4 \times 10^{-53}$

$(s)(s) = 4 \times 10^{-53}$

$s = 6 \times 10^{-27}$ $\underline{M}$

Bi_2S_3 is 4×10^7 times more soluble than HgS!

36. No more than 0.1% of the Ba^{2+} must remain, or < 0.20 mg (in 100 mL).

After precipitation,

$[Ba^{2+}] = (2.0 \times 10^{-3}$ mg/mL)/(137 mg/mmol) $= 1.5 \times 10^{-5}$ $\underline{M}$

$[Ba^{2+}][F^-]^2 = 1.7 \times 10^{-6}$

$(1.5 \times 10^{-5})[F^-]^2 = 1.7 \times 10^{-6}$

$[F^-] = 0.33$ $\underline{M}$

This concentration of excess fluoride is attainable, so the analysis in principle would work.

37. (a) $K_{sp} = a_{Ba^{2+}} \cdot a_{SO_4^{2-}} = [Ba^{2+}]f_{Ba^{2+}} \cdot [SO_4^{2-}]f_{SO_4^{2-}} = K_{sp}f_{Ba^{2+}}f_{SO_4^{2-}}$

(b) $K_{sp} = a_{Ag^{2+}} \cdot a_{CrO_4^{2-}} = [Ag^+]^2f_{Ag^{2+}} \cdot [CrO_4^{2-}]f_{CrO_4^{2-}} = K_{sp}f_{Ag^{2+}}f_{CrO_4^{2-}}$

38. $\mu = 0.0375$. From a calculation using Equation 4.19, $f_{Ba^{2+}} = 0.502$, $f_{SO_4^{2-}} =$ 0.485. From the appendix, K_{sp} at zero ionic strength ($K_{sp}°$) is 1.0×10^{-10}.

$K_{sp} = [Ba^{2+}][SO_4^{2-}]f_{Ba^{2+}}f_{SO_4^{2-}}$

$1.0 \times 10^{-10} = (s)(s)(0.502)(0.485)$

$s = 2.0 \times 10^{-5}$ $\underline{M}$

39. $CaF_2 = Ca^{2+} + 2F^-$

$0.015 + 1/2x \quad x$

≈ 0.015

$K_{sp}° = [Ca^{2+}]\, f_{Ca^{2+}} \cdot [F^-]^2 f_{F^-}^{\,2} = K_{sp} f_{Ca^{2+}} f_{F^-}^{\,2}$

$K_{sp} = K_{sp}°/f_{Ca^{2+}} f_{F^-}^{\,2} = [Ca^{2+}][F^-]^2$

From the appendix, $K_{sp}° = 4.0 \times 10^{-11}$

The solution contains 0.015 $\underline{M}$ $Ca(NO_3)_2$ + 0.025 $\underline{M}$ $NaNO_3$ (neglect amount of F_2 in solution).

$\mu = 1/2\ ([Ca^{2+}](2)^2 + [Na^+](1)^2 + [NO_3^-](1)^2)$

$= 1/2\ [(0.015)(4) + (0.025)(1) + (0.040)(1)] = 0.062$

From Reference 9 in Chapter 6,

$\alpha_{Ca^{2+}} = 6 \quad \alpha_{F^-} = 3.5$

From Equation 6.19 in Chapter 6,

$f_{Ca^{2+}} = 0.46$; $f_{F^-} = 0.76$

$K_{sp} = (4.0 \times 10^{-11})/(0.46)(0.76) = 1.1_4 \times 10^{-10}$

$(0.015)(x)^2 = 1.1_4 \times 10^{-10}$

$x = 8.7 \times 10^{-5}$ $\underline{M}$ $= [F^-]$

$(8.7 \times 10^{-5}\ mol/L)(0.25\ L)(19.0\ g/mol) = 4.1 \times 10^{-4}$ g F^- in solution

40. See the CD for the spreadsheet. The calculated value is 16.298% P_2O_5. For a 0.5267 g sample and 2.0267 g precipitate, the %P_2O_5 is 14.55_4%.

41. See the CD for the spreadsheet and graph.

42. See the CD for the spreadsheet and graph.

43. See the CD for the spreadsheet calculation.

1. *The Volhard titration involves adding excess standard silver nitrate solution to a chloride solution and then back titrating the excess silver with standard potassium thiocyanate solution. The AgCl precipitate is first removed by filtration, or else nitrobenzene is added to prevent its reaction with SCN^-. The indicator is Fe^{3+} and the end point is marked by formation of the red $Fe(SCN)^{2+}$ complex. The Fajans titration involves the direct titration of chloride. The end point is marked by adsorption of the indicator, dichlorofluorescein, which causes it to turn pink. Dextrin is added to minimize coagulation at the end point. The Volhard method must be used for strong acid solution, because the adsorption indicator in the Fajan titration is too slightly dissociated in acid solution for the anion form to be adsorbed.*
2. *The ionized form of the indicator, which has a charge opposite that of the titrating ion, becomes adsorbed on the surface of the precipitate with the first drop of excess titrant. The excess titrant imparts a charge to the precipitate surface, which attracts the indicator ion. The adsorbed indicator has a color distinctly different from the nonadsorbed indicator, possibly due to complexation with the titrant in the precipitate.*

3. From the appendix, K_{sp} of $AgIO_3 = 3.1 \times 10^{-8}$ and K_a of $HIO_3 = 2 \times 10^{-1}$. The equilibria are

$$AgIO_3 = Ag^+ + IO_3^- \quad (1)$$

$$IO_3^- + H^+ = HIO_3 \quad (2)$$

$$K_{sp} = [Ag^+][IO_3^-] = [Ag^+]C_{HIO_3}\alpha_1 \quad (3)$$

where α_1 is the fraction of the total iodate (C_{HIO_3}) that exists as IO_3^-. Then,

$$K_{sp}/\alpha_1 = K_{sp}' = [Ag^+]C_{HIO_3} = s^2 \quad (4)$$

where s is the water solubility. First calculate α_1.

$$C_{HIO_3} = [HIO_3] + [IO_3^-] \quad (5)$$

But, $[HIO_3] = [H^+][IO_3^-]/K_a$. So,

$$C_{HIO_3} = [H^+][IO_3^-]/K_a + [IO_3^-] \quad (6)$$

from which

$$1/\alpha_1 = C_{HIO_3}/[IO_3^-] = [H^+]/K_a + 1 = 1.5 \text{ (at 0.100 } \underline{M} \text{ acid)}$$

$$K_{sp}' = (3.1 \times 10^{-8})(1.5) = s^2$$

$$s = 2.1 \times 10^{-4}\ \underline{M} = [Ag^+] = C_{HIO_3}$$

(This is only increased from 1.8×10^{-4} $\underline{M}$ in the absence of acid, since HIO_3 is a fairly strong acid.)

$$\alpha_1 = 0.67;\ \alpha_0 = 1 - 0.67 = 0.33$$

$$\therefore [IO_3^-] = (2.1 \times 10^{-4})(0.67) = 1.5 \times 10^{-4}\ \underline{M}$$

$[HIO_3] = (2.1 \times 10^{-4})(0.33) = 6.9 \times 10^{-5}$ M

4. From the appendix, K_{sp} of $CaF_2 = 4.0 \times 10^{-11}$ and K_a of HF $= 6.7 \times 10^{-4}$. The equilibria are

$CaF_2 = Ca^{2+} + 2F^-$ (1)

$F^- + H^+ = HF$ (2)

$K_{sp} = [Ca^{2+}][F^-]^2 = [Ca^{2+}]C_{HF}^{\,2}\alpha_1^{\,2}$ (3)

(since $[F^-] = C_{HF}\alpha_1$).

$K_{sp}/\alpha_1^{\,2} = K_{sp}' = [Ca^{2+}]C_{HF}^{\,2} = (s)(2s)^2$ (4)

As in the preceding problem,

$1/\alpha_1 = [H^+]/K_a + 1 = 1.5_1 \times 10^2$ (At 0.100 M acid)

$K_{sp}' = (4.0 \times 10^{-11})(1.5_1 \times 10^2)^2 = (s)(2s)^2$

$s = 6.1 \times 10^{-3}$ M $= [Ca^{2+}] = 1/2\ C_{HF}$

$\alpha_1 = 6.6_2 \times 10^{-3}$; $\alpha_0 = 1 - 6.6_2 \times 10^{-3} = 0.993$

Since $C_{HF} = 2 \times 6.1 \times 10^{-3} = 1.2_1 \times 10^{-2}$ M,

$[HF] = (1.2_1 \times 10^{-2})(0.993) = 1.2_0 \times 10^{-2}$ M

$[F^-] = (1.2_1 \times 10^{-2})(6.6_2 \times 10^{-3}) = 8.0_1 \times 10^{-5}$ M

Essentially all of the fluoride exists as HF and its concentration is twice that of the calcium. About 12% of the HCl was consumed in forming HF. Recalculation (reiteration) using 0.088 M HCl would result in a minor correction in the calculation ($s = 5.6 \times 10^{-3}$ M).

5. From the appendix, K_{sp} of PbS $= 8 \times 10^{-28}$ and $K_{a_1} = 9.1 \times 10^{-8}$, $K_{a_2} = 1.2 \times 10^{-15}$ for H_2S. The equilibria are:

$PbS = Pb^{2+} + S^{2-}$ (1)

$S^{2-} + H^+ = HS^-$ (2)

$HS^- + H^+ = H_2S$ (3)

$$K_{sp} = [Pb^{2+}][S^{2-}] = [Pb^{2+}]C_{H_2S}\alpha_2 \quad (4)$$

$$K_{sp}/\alpha_2 = K_{sp}' = [Pb^{2+}]C_{H_2S} = s^2 \quad (5)$$

We calculate for the diprotic acid that

$$\alpha_0 = ([H^+]^2)/([H^+]^2 + K_{a_1}[H^+] + K_1K_{a_2}) = 1.00 \quad \text{(for 0.0100 } \underline{M} \text{ acid)}$$

$$\alpha_1 = (K_{a_1}[H^+])/([H^+]^2 + K_{a_1}[H^+] + K_{a_1}K_{a_2}) = 9.1 \times 10^{-6}$$

$$\alpha_2 = (K_{a_1}K_{a_2})/([H^+]^2 + K_{a_1}[H^+] + K_{a_1}K_{a_2}) = 1.0_9 \times 10^{-18}$$

$$K_{sp}' = (8 \times 10^{-28})/(1.0_9 \times 10^{-18}) = s^2$$

$$s = 2._7 \times 10^{-5}\ \underline{M} = [Pb^{2+}] = C_{H_2S}$$

$$[H_2S] = C_{H_2S}\alpha_0 = (2._7 \times 10^{-5})(1.00) = 2._7 \times 10^{-5}\ \underline{M}$$

$$[HS^-] = C_{H_2S}\alpha_1 = (2._7 \times 10^{-5})(9.1 \times 10^{-6}) = 2.5 \times 10^{-10}\ \underline{M}$$

$$[S^{2-}] = C_{H_2S}\alpha_2 = (2._7 \times 10^{-5})(1.0_9 \times 10^{-18}) = 2._9 \times 10^{-23}\ \underline{M}$$

Virtually all the sulfide is in the H_2S form.

6. From the appendix, K_{sp} of AgCl is 1.0×10^{-10}. The equilibria are:

$$AgCl = Ag^+ + Cl^- \quad (1)$$

$$Ag^+ + en = Ag(en)^+ \quad (2)$$

$$Ag(en)^+ + en = Ag(en)_2^+ \quad (3)$$

$$K_{sp} = [Ag^+][Cl^-] = C_{Ag}\beta_0\ [Cl^-] \quad (4)$$

$$K_{sp}/\beta_0 = K_{sp}' = C_{Ag}[Cl^-] = s^2 \quad (5)$$

For a complex with two ligands [similar to $Ag(NH_3)_2^+$ -- see Equations 9.20–9.22], we calculate that

$$\beta_0 = 1/(K_{f_1}K_{f_2}[en]^2 + K_{f_1}[en] + 1) = 1.4_3 \times 10^{-6}. \quad \text{(for 0.100 } \underline{M} \text{ en)}$$

$$\beta_1 = (K_{f_1}[en])/(K_{f_1}K_{f_2}[en]^2 + K_f[en] + 1) = 7.1 \times 10^{-3}$$

$\beta_2 = (K_{f_1}K_{f_2}[en]^2)/(K_{f_1}K_{f_2}[en]^2 + K_{f_1}[en] + 1) = 1.00$

$K_{sp}' = (1.0 \times 10^{-10})/(1.4_3 \times 10^{-6}) = s^2$

$s = 8.4 \times 10^{-3}\ \underline{M} = C_{Ag} = [en]$

$[Ag^+] = C_{Ag}\beta_0 = (8.4 \times 10^{-3})(1.4_3 \times 10^{-6}) = 1.2_0 \times 10^{-8}\ \underline{M}$

$[Ag(en)^+] = C_{Ag}\beta_1 = (8.4 \times 10^{-3})(7.1 \times 10^{-3}) = 5.9_6 \times 10^{-5}\ \underline{M}$

$[Ag(en)_2^+] = C_{Ag}\beta_2 = (8.4 \times 10^{-3})(1.00) = 8.4 \times 10^{-3}\ \underline{M}$

7. *The equilibria are:*

$AgIO_3 \rightleftharpoons Ag^+ + IO_3^-$

$IO_3^- + H^+ \rightleftharpoons HIO_3$

$H_2O \rightleftharpoons H^+ + OH^-$

$HNO_3 \rightleftharpoons H^+ + NO_3^-$

The equilibrium constant expressions are:

$K_{sp} = [Ag^+][IO_3^-] = 3.1 \times 10^{-8}$ (1)

$K_a = [H^+][IO_3^-]/[HIO_3] = 2 \times 10^{-1}$ (2)

$K_w = [H^+][OH^-] = 1.00 \times 10^{-14}$ (3)

The mass balance expressions are:

$[Ag^+] = [IO_3^-] + [HIO_3] = C_{HIO_3}$ (4)

$[H^+] = [NO_3^-] + [OH^-] - [HIO_3]$ (5)

$[NO_3^-] = 0.100\ \underline{M}$ (6)

The charge balance expression is:

$[H^+] + [Ag^+] = [IO_3^-] + [NO_3^-] + [OH^-]$ (7)

There are six unknowns ($[H^+]$, $[OH^-]$, $[NO_3^-]$, $[HIO_3]$, $[IO_3^-]$, $[Ag^+]$) and 6 independent equations.

Simplifying assumptions:

(1) Assume $[OH^-]$ is very small.

(2) Assume $[HIO_3]$ is small compared to $[H^+]$, since K_{sp} is small.

(3) K_a is large, but so is the acidity. Let's assume, though, that $[IO_3^-] >> [HIO_3]$, i.e., $\alpha_1 >> \alpha_2$ (it is actually 67% of the total -- see Problem 3).

With these assumptions, from (5) and (6),

$$[H^+] = 0.100 + [OH^-] - [HIO_3] \approx 0.100 \quad (8)$$

From (1)

$$[Ag^+] = K_{sp}/[IO_3^-] \quad (9)$$

From assumption (3), $[Ag^+] \approx [IO_3^-]$

$$[Ag^+] = K_{sp}/[Ag^+] = 3.1 \times 10^{-8}/[Ag^+] \quad (10)$$

$[Ag] = 1.8 \times 10^{-4}$ M

This compares with 2.1×10^{-4} M calculated in Problem 3, being 14% low because of the assumption that HIO_3 does not form (which increases the solubility). To be more correct, exact solution of the simultaneous equations is needed, or we can calculate α_1 (it is 0.67, so $[HIO_3] = 0.33\,[Ag^+]$ and following from (9) and (2),

$[Ag^+] = K_{sp}[H^+]/K_a[HIO_3] = K_{sp}[H^+]/K_a\, 0.33[Ag^+]$

$= (3.1 \times 10^{-8})(0.100)/(2 \times 10^{-1})(0.33)[Ag^+]$

$[Ag^+] = 2.1 \times 10^{-4}$ M

The same is obtained by substituting $0.67\,[Ag^+]$ for $[IO_3^-]$ in (9).

8. *Equilibria:*

$PbS \rightleftharpoons Pb^{2+} + S^{2-}$

$S^{2-} + H^{+} \rightleftharpoons HS^{-}$

$HS^{-} + H^{+} \rightleftharpoons H_2S$

Equilibrium expressions:

$[Pb^{2+}][S^{2-}] = K_{sp} = 8 \times 10^{-28}$ (1)

$[H^{+}][HS^{-}] / [H_2S] = K_{a1} = 9.1 \times 10^{-8}$ (2)

$[H^{+}][S^{2-}] / [HS^{-}] = K_{a2} = 1.2 \times 10^{-15}$ (3)

Mass balance expressions:

$[Pb^{2+}] = [S^{2-}] + [HS^{-}] + [H_2S]$ (4)

$[H^{+}] = [Cl^{-}] + [OH^{-}] - [HS^{-}] - [H_2S]$ (5)

$[Cl^{-}] = 0.0100 \underline{M}$ (6)

6 unknowns ($[Pb^{2+}]$, $[S^{2-}]$, $[HS^{-}]$, $[H_2S]$, $[H^{+}]$, $[OH^{-}]$) and 6 independent equations

Assume: $[H^{+}] >> [OH^{-}]$, $[HS^{-}]$, and $[H_2S]$

and $[H_2S] >> [HS^{-}]$ and $[S^{2-}]$

Then, from (4)

$[Pb^{2+}] \approx [H_2S]$

From (5),

$[H^{+}] \approx 0.0100 \underline{M}$

Calculate $[Pb^{2+}]$:

From (1), $[Pb^{2+}] = K_{sp}/[S^{2-}]$ (7)

From (3), $[S^{2-}] = K_{a2}[HS^{-}]/[H^{+}]$ (8)

From (2), $[HS^{-}] = K_{a1}[H_2S]/[H^{+}] \approx K_{a1}[Pb^{2+}]/[H^{+}]$ (9)

$\therefore$ $[S^{2-}] = K_{a1}K_{a2}[Pb^{2+}]/[H^{+}]^2$ (10)

Substituting (10) in (7):

$[Pb^{2+}] = K_{sp}[H^+]^2/K_{a1}K_{a2}[Pb^{2+}]$

$= (8 \times 10^{-28})(1.00 \times 10^{-2})^2/(9.1 \times 10^{-8})(1.2 \times 10^{-15})[Pb^{2+}] = 7.3 \times 10^{-10}/[Pb^{2+}]$

Solving, $[Pb^{2+}] = 2.7 \times 10^{-5}$ $\underline{M}$

This is the same answer as calculated in Problem 5.

9. *Equilibria*

$Ag^+ + en = Ag(en)^+$	$K_{f1} = [Ag(en)^+]/[Ag^+][en] = 5.0 \times 10^4$	(1)
$Ag(en)^+ + en = Ag(en)_2{}^+$	$K_{f2} = [Ag(en)_2{}^+]/[Ag(en)^+][en] = 1.4 \times 10^3$	(2)
$AgCl = Ag^+ + Cl^-$	$K_{sp} = [Ag^+][Cl^-] = 1.0 \times 10^{-10}$	(3)

Mass balance expressions:

$[en] = 0.100\ M - [Ag(en)^+] - [Ag(en)_2{}^+]$ (4)

$[Ag^+] = [Cl^-] - [Ag(en)^+] - [Ag(en)_2{}^+]$ (5)

or

$[Cl^-] = [Ag^+] + [Ag(en)^+] + [Ag(en)_2{}^+]$

Charge balance expression:

$[Ag^+] + [Ag(en)^+] + [Ag(en)_2{}^+] = [Cl^-]$ (6) ≡ (5)

5 unknowns: $[Ag^+]$, $[Ag(en)^+]$, $[Ag(en)_2{}^+]$, $[en]$, $[Cl^-]$

5 independent equations

Simplifying assumptions:

Since K_{f1} *and* K_{f2} *are fairly large, assume* $[Ag(en)_2{}^+] >> [Ag(en)^+]$ *and* $[Ag^+]$

∴ $[Cl^-] \approx [Ag(en)_2{}^+]$ = *solubility of AgCl = s*

Assume solubility is small, and so $[en] \approx 0.100$ $\underline{M}$

From (3), $[Cl^-] = K_{sp}/[Ag^+] = s$ (7)

From (1), $[Ag^+] = [Ag(en)^+]/K_{f1}[en]$ (8)

From (2), $[Ag(en)^+] = [Ag(en)_2{}^+]/K_{f2}[en]$ (9)

Substituting (9) in (8),

$[Ag^+] = [Ag(en)_2^+]/K_{f1}K_{f2}[en]^2 \approx [Cl^-]/K_{f1}K_{f2}[en]^2$ (10)

$\therefore \quad [Cl^-] = K_{sp}K_{f1}K_{f2}[en]^2/[Cl^-]$

$[Cl^-] = \sqrt{K_{sp}K_{f1}K_{f2}[en]^2} = \sqrt{(1.0 \times 10^{-10})(5.0 \times 10^4)(1.4 \times 10^3)(0.100)^2}$

$= 8.4 \times 10^{-3}\ \underline{M} = s$

This is the same solubility calculated in Problem 6.

Note that in both Problem 6 and this one, we neglected the amount of en consumed (= 2 x 8.4 x 10^{-3} = 0.016_8 $\underline{M}$, leaving 0.083_2 $\underline{M}$). We could reiterate either problem using this new concentration of en, giving s = 7.0 x 10^{-3} $\underline{M}$.

10. mg Cl in sample = (0.1182$\underline{M}$ x 15.00 mL - 0.101 $\underline{M}$ x 2.38 mL) x 35.45 mg/mmol

= 54.34 mg

(54.34 mg/10.00 mL) x 10^3 mL/L x 10^{-3} g/mg = 5.434 g/L

11. In order for Ag_2CrO_4 to precipitate, the Ag^+ concentration is given by

$[Ag^+]^2(0.0011) = 1.1 \times 10^{-12}$

$[Ag^+] = 3.2 \times 10^{-5}\ \underline{M}$

This must come from the solubility of AgCl plus excess titrant. Calculate the solubility of AgCl in the presence of 3.2×10^{-5} $\underline{M}$ Ag^+.

$AgCl = Ag^+ + Cl^-$

$\quad\quad 3.2 \times 10^{-5} \quad s$

$(3.2 \times 10^{-5})(s) = 1.0 \times 10^{-10}$

s = 3.1×10^{-6} $\underline{M}$ = $[Ag^+]$ produced from the precipitate.

Total mmol Ag^+ = 3.2×10^{-5} $\underline{M}$ x 100 mL = 3.2×10^{-3}

mmol Ag^+ from ppt. = 3.1×10^{-6} $\underline{M}$ x 100 mL = 0.31×10^{-3}

mmol excess Ag^+ from titrant = 2.9×10^{-3}

0.100 $\underline{M}$ x x mL = 2.9×10^{-3} mmol

x = 0.029 mL excess titrant

CHAPTER 11

12. *See CD for spreadsheet and plot of titration curve.*

1. *An oxidizing agent takes on electrons from a reducing agent. The former is reduced to a lower valence state and the latter is oxidized to a higher oxidation state.*
2. *The Nernst equation defines the potential of an electrode in a solution containing a redox couple:* $aOx + ne^- = bRed$

   $E = E^o - (2.303\ RT/nF)\ \log([Red]^b/[Ox]^a)$
3. *The standard potential (E^o) is the potential of an electrode in a solution relative to the normal hydrogen electrode, with all species at unit activity. The formal potential ($E^{o\prime}$) is the potential of an electrode under specified solution conditions.*
4. *A salt bridge prevents mixing of two solutions but allows charge transfer between them.*
5. *Normal hydrogen electrode (N.H.E.) or standard hydrogen electrode (S.H.E.). The standard potential of the half reaction* $2H^+ + 2e^- = H_2$ *is arbitrarily defined as zero and all other potentials are referred to this.*
6. *A poor reducing agent.*
7. *About 0.2—0.3 V*
8. *They predict whether there is sufficient driving force (i.e., potential) for the reaction, but they say nothing of the rate of the reaction.*
9. O_3, $HClO$, Hg^{2+}, H_2SeO_3, H_3AsO_4, Cu^{2+}, Co^{2+}, Zn^{2+}, K^+
10. Ni, H_2S, Sn^{2+}, V^{3+}, I^-, Ag, Cl^-, Co^{2+}, HF

11. (a) $Fe^{2+} - MnO_4^-$

(b) $Fe^{2+} - Ce^{4+}(HClO_4)$

(c) $H_3AsO_3 - MnO_4^-$

(d) $Fe^{3+} - Ti^{2+}$

12. (a) $Pt/Fe^{2+}, Fe^{3+}//Cr_2O_7^{2-}, Cr^{3+}, H^+/Pt$

(b) $Pt/I^-, I_2//IO_3^-, I_2, H^+/Pt$

(c) $Zn/Zn^{2+}//Cu^{2+}/Cu$

(d) $Pt/H_2SeO_3, SeO_4^{2-}, H^+//Cl_2, Cl^-/Pt$

13. (a) $2V^{2+} + PtCl_6^{2-} = 2V^{3+} + PtCl_4^{2-} + 2Cl^-$

(b) $Ag + Fe^{3+} + Cl^- = \underline{AgCl} + Fe^{2+}$

(c) $3Cd + ClO_3^- + 6H^+ = 3Cd^{2+} + Cl^- + 3H_2O$

(d) $2I^- + H_2O_2 + 2H^+ = I_2 + 2H_2O$

14. $E = 1.52 - (0.059/5) \log [Br_2]^{1/2}/([BrO_3^-][H^+]^6)$

$[H^+] = 10^{-2.5} = 10^{.5} \times 10^{-3} = 3.2 \times 10^{-3}\ \underline{M}$

$E = 1.52 - (0.059/5) \log [(0.20)^{1/2}/(0.50)(3.2 \times 10^{-3})^6] = 1.34\ V$

15. Since I^- is in large excess, I_3^- is formed instead of I_2:

$H_2O_2 + 3I^- + 2H^+ = 2H_2O + I_3^-$

Since both $[I^-]$ and $[I_3^-]$ are known, use this couple for calculations.

mmol I_3^- formed = mmol H_2O_2 at start = 0.10 x 10 = 1.0 mmol

$[I_3^-]$ = 1.0 mmol/100 mL = 0.010 $\underline{M}$

$[I^-] = [(5.0 \times 90) - 3 \times 1.0]$ mmol/100 mL = 4.5 $\underline{M}$

$E = 0.536 - (0.059/2) \log ([I^-]^3/[I_3^-])$

$= 0.536 - (0.059/2) \log [(4.5)^3/(0.010)] = 0.419$ V

16. $E = 0.68 - (0.059/2) \log ([PtCl_4^{2-}][Cl^-]^2/[PtCl_6^{2-}])$

We can assume $[Cl^-] = 3.0$ $\underline{M}$ since the amount used in complexing the Pt is small.

$\therefore E = 0.68 - (0.059/2) \log [(0.025)(3.0)^2/(0.015)] = 0.65$ V

17. $UO_2^{2+} + 4H^+ + 2e^- = U^{4+} + 2H_2O \qquad E^0 = 0.334$ V

$V^{3+} + e^- = V^{2+} \qquad E^0 = -0.255$ V

$UO_2^{2+} + 2V^{2+} + 4H^+ = U^{4+} + 2V^{3+} + 2H_2O \qquad E^0_{cell} = 0.589$ V

Assuming 1 mL each, we have 0.1 mmol of UO_2^{2+}, 0.1 mmol of V^{2+}, and 0.2 mmol H_2SO_4 (0.4 mmol H^+) in 2 mL. Each millimole of V^{2+} will react with only 1/2 mmol of UO_2^{2+} and 2 mmol H^+, so we have an excess of UO_2^{2+}.

mmol UO_2^{2+} = 0.100 - 0.050 = 0.050 mmol left/2 mL

mmol U^{4+} produced = 0.050 mmol/2 mL

mmol H^+ = 0.40 - 0.20 = 0.20 mmol left/2 mL = 0.10 $\underline{M}$

Volumes cancel for the uranium species, so we can use millimoles:

$\therefore E = 0.334 - (0.059/2) \log ([U^{4+}]/[UO_2^{2+}][H^+]^4)$

$= 0.334 - (0.059/2) \log [(0.050)/(0.050)(0.10)^4] = 0.216$ V

18. Since the $PtCl_6^{2-}/PtCl_4^{2-}$ potential is the more positive, $PtCl_6^{2-}$ will oxidize V^{2+}, or subtracting the second half-reaction (multiplied by 2) from the first to give a positive E_{cell}, the reaction is:

$PtCl_6^{2-} + 2V^{2+} = PtCl_4^{2-} + 2V^{3+} + 2Cl^-$ $E_{cell} = 0.68 - (-0.255)\ V = 0.94\ V$

19. (a) $E_{cell} = E_{cathode} - E_{anode} = 1.20 - (0.059/10) \log ([I_2]/[IO_3^-]^2[H^+]^{12}) - 0.5355 + (0.059/2) \log ([I^-]^3/[I_3^-])$

$E_{cell} = 1.20 - (0.059/10) \log [(0.0100)/(0.100)(0.100)^{12}]$
$- 0.54 + (0.059/2) \log [(0.100)^3/(0.0100)] = 0.57\ V$

(b) $E_{cell} = 0.334 - (0.059/2) \log ([U^{4+}]/[UO_2^{2+}][H^+]^4)$
$- 0.222 + 0.059 \log [Cl^-]$

$E_{cell} = 0.334 - (0.059/2) \log [(0.0500)/(0.200)(1.00)^4]$
$- 0.222 + 0.059 \log (0.100) = 0.071\ V$

(c) $E_{cell} = 1.51 - (0.059/5) \log ([Mn^{2+}]/[MnO_4^-][H^+]^8)$
$- 1.25 + (0.059/2) \log ([Tl^+]/[Tl^{3+}])$

$E_{cell} = 1.51 - (0.059/5) \log [(0.100)/(0.0100)(1.0 \times 10^{-2})^8]$
$- 1.25 + (0.059/2) \log (0.100/0.0100) = 0.09\ V$

20. $VO_2^+ + 2H^+ + e^- = VO^{2+} + H_2O$ $E° = 1.00\ V$ (1)

$UO_2^{2+} + 4H^+ + 2e^- = U^{4+} + 2H_2O$ $E° = 0.334\ V$ (2)

Subtracting (2) from (1) (multiplied by 2) gives a positive cell potential and hence, the spontaneous reaction:

$2VO_2^+ + U^{4+} = 2VO^{2+} + UO_2^{2+}$ $E°_{cell} = 1.00 - 0.334 = 0.67\ V$

1. *The liquid junction potential is the potential existing at a boundary between two dissimilar solutions (e.g., at a solution-salt bridge interface). It is due to the unequal diffusion of ions across the boundary, which results in a net charge built up on each side of the boundary (a potential). It can be minimized by adding a high concentration of an electrolyte on one side of the boundary, whose cation and anion diffuse at nearly equal rates (e.g., saturated KCl).*

   The residual liquid junction potential is the difference in liquid junction potentials of an electrode in a calibrating solution and an unknown solution. It can be minimized by keeping the conditions of the solutions as nearly the same as possible, such as the ionic strength, and especially the pH.

2. *The generally accepted theory of the glass electrode response to protons is the result of migration of solution protons to the hydrated glass surface gel containing low activity protons from the –SiOH groups, building up a microscopic layer of positive charges, the boundary potential. Other theories are offered based on chemisorption and charge separation (Pungor) or capacitance (Cheng).*

3. *The alkaline error is the result of competition between H^+ and Na^+ ions for the potential determining mechanism. At high pH, response to Na^+ becomes appreciable, causing the pH reading to be "too acidic". The acid error occurs in very acidic solution and is a result of the decrease of water activity, causing a positive error in the pH reading.*

4. *The general construction of ion selective electrodes is: internal reference electrode/internal filling solution/membrane. The main difference in the electrodes lies in differences in their membranes. The four major types of membranes include glass*

membranes (useful for monovalent cations), precipitate impregnated membranes (useful for anions and some cations), solid state (crystalline) membranes (useful for halides, especially fluoride, sulfide, cyanide, and some cations), and liquid-liquid membranes (useful for several cations and anions, especially Ca^{2+}, nitrate, perchlorate, and chloride).

5. *It is a measure of the selectivity of the membrane response of an ion selective electrode for one ion over another and can be used in the Nernst equation to predict the response due to an interfering ion. It can be estimated by measuring the potential of two different solutions containing different concentrations of the two ions to which the electrode responds (separate solution method), or by measuring the potential of solution of mixtures of the ions (mixed solution method).*

6. *A crown ether is a neutral cyclic ether containing several oxygens in a ring or cage that is of appropriate size to incorporate and complex metal ions of certain sizes. A 16-crown-6 ether contains 6 oxygens in a 16-membered ring.*

7. *The Nicolsky equation describes the potential of an electrode that responds to two (or more) ions, and is given by Equation 13.49.*

8. $E^\circ_{AgBr,Ag} = E^\circ_{Ag^+,Ag} + 0.05916 \log K_{sp}$

$0.073 = 0.799 + 0.05916 \log K_{sp}$

$\log K_{sp} = -12.2_7$

$K_{sp} = 10^{-12.2_7} = 10^{-13} \times 10^{0.7_3} = 5._4 \times 10^{-13}$

9. $E_{cell} = E^\circ_{Ag^+,Ag} - 0.05916 \log (1/a_{Ag^+}) - E_{S.C.E.}$

At the end point, a saturated solution of AgSCN exists.

$\therefore [Ag^+] = \sqrt{K_{sp}} = 1.00 \times 10^{-6}$ M

$0.202 = E°_{Ag^+,Ag} - 0.05916 \log [1/(1.00 \times 10^{-6})] - 0.242$

$E°_{Ag^+,Ag} = 0.799$ V

10. (a) (1) $Ag = Ag^+ + e^-$ $E° = 0.799$

$Fe^{3+} + e^- = Fe^{2+}$ $E° = 0.771$

(2) $Ag/Ag^+, Fe^{3+}, Fe^{2+}/Pt$

(3) $E°_{cell} = E°_{right} - E°_{left} = 0.771 - 0.799 = -0.028$ V

Since the potential is negative, the reaction will not go as written under standard conditions.

(4) Ag electrode = -; Pt electrode = +

(b) (1) $VO_2^+ + 2H^+ + e^- = VO^{2+} + H_2O$ $E° = 1.000$

$V^{3+} + H_2O = VO^{2+} + 2H^+ + e^-$ $E° = 0.361$

(2) $Pt/V^{3+}, VO_2^+, VO^{2+}/Pt$

(3) $E°_{cell} = 1.000 - 0.361 = 0.639$ V

(4) left Pt electrode = -; right Pt electrode = +

(c) (1) $Ce^{4+} + e^- = Ce^{3+}$ $E° = 1.61$

$Fe^{2+} = Fe^{3+} + e^-$ $E° = 0.771$

(2) $Pt/Fe^{2+}, Fe^{3+}, Ce^{4+}, Ce^{3+}/Pt$

(3) $E°_{cell} = 1.61 - 0.77 = 0.84$ V

(4) Left Pt electrode = -; right Pt electrode = +

11. (a) (1) $H_2 = 2H^+ + 2e^-$ $E° = 0.000\ V$

$Cl_2 + 2e^- = 2Cl^-$ $E° = 1.359\ V$

$Cl_2 + H_2 = 2HCl$

(2) $E_{cell} = E_{right} - E_{left}$

$= 1.359 - (0.05916/2) \log ([Cl^-]^2/p_{Cl_2}) + (0.05916/2)$

$\log (p_{H_2}/[H^+]^2) = 1.359 - (0.05916/2) \log [(0.5)^2/(0.2)]$

$+ (0.05916/2) \log [(0.2)/(0.5)^2] = 1.353\ V$

(b) (1) $Fe^{2+} = Fe^{3+} + e^-$ $E° = 0.771\ V$

$VO_2^+ + 2H^+ + e^- = VO^{2+} + H_2O$ $E° = 1.000\ V$

$VO_2^+ + Fe^{2+} + 2H^+ = VO^{2+} + Fe^{3+} + H_2O$

(2) $E_{cell} = 1.000 - 0.5916 \log [VO^{2+}]/(VO_2^+][H^+]^2)$

$- 0.771 + 0.05916 \log ([Fe^{2+}]/[Fe^{3+}])$

$= 1.000 - 0.05916 \log (0.002)/[(0.001)(0.1)^2]$

$- 0.771 + 0.05916 \log [(0.005)/(0.05)] = 0.034\ V$

12. (a) $E_{vs\ SCE} = 1.087 - 0.242 = 0.845\ V$

(b) $E_{vs\ SCE} = 0.222 - 0.242 = -0.020\ V$

(c) $E_{vs\ SCE} = -0.255 - 0.242 = -0.497\ V$

13. $E_{Pt} = 0.771 - 0.05916 \log [(0.0500)/(0.00200)] = 0.688\ V$

$E_{cell} = E_{Pt} - E_{SCE} = 0.688 - 0.242 = 0.446\ V$

14. (a) When the I^- is all titrated, the volume is doubled and $[Cl^-] = 0.050\ \underline{M}$ when AgCl begins to precipitate:

$[Ag^+] = K_{sp(AgCl)}/[Cl^-] = (1.0 \times 10^{-10})/(0.050) = 2.0 \times 10^{-9}\ \underline{M}$

$[I^-] = K_{sp(AgCl)}/[Ag^+] = (1 \times 10^{-16})/(2.0 \times 10^{-9}) = 5 \times 10^{-8}\ \underline{M}$

% I^- remaining = $[(5 \times 10^{-8}\ \underline{M} \times 100\ mL)/(1.0 \times 10^{-1}\ \underline{M} \times 50\ mL)] \times 100\%$

$= 1 \times 10^{-4}\%$

(b) $E = E°_{Ag^+,Ag} - 0.05916 \log (1/a_{Ag^+}) - E_{S.C.E.}$

$- 0.799 - 0.05916 \log [1/(2.0 \times 10^{-9})] - 0.242 = 0.042$ V

At the theoretical end point, we have a saturated solution of AgI, and:

$[Ag^+] = \sqrt{K_{sp(AgI)}} = \sqrt{1 \times 10^{-16}} = 1 \times 10^{-8}\ \underline{M}$

$E = 0.799 - 0.05916 \log [1/(1 \times 10^{-8})] - 0.242 = 0.08_4$ V

(c) At the end point, we have a saturated solution of AgCl, and:

$[Ag^+] = \sqrt{K_{sp(AgCl)}} = \sqrt{1.0 \times 10^{-10}} = 1.0 \times 10^{-5}\ \underline{M}$

$E = 0.799 - 0.05916 \log [1/(1.0 \times 10^{-5})] - 0.242 = 0.261$ V

15. $E = E°_{Hg^{2+},Hg} - (2.303\ RT/2F) \log (1/a_{Hg^{2+}})$

$K_{f(Hg-EDTA)} = (a_{Hg-EDTA})/(a_{Hg^{2+}} \cdot a_{EDTA^{2-}})$

$a_{Hg^{2+}} = [1/K_{f(Hg-EDTA)}] \cdot (a_{Hg-EDTA}/a_{EDTA^{2-}})$

Substituting this in the equation,

$E = E°_{Hg^{2+},Hg} - (2.303\ RT/2F) \log K_{f(Hg-EDTA)} - (2.303\ RT/2F$

$\log (a_{EDTA^{2-}}/a_{Hg-EDTA})$

$K_{f(M-EDTA)} = (a_{M-EDTA})/(a_{M^{n+}} \cdot a_{EDTA^{2-}})$

$a_{EDTA^{2-}} = [1/K_{f(M-EDTA)}]\ (a_{M-EDTA}/a_{M^{n+}})$

Substituting this in the equation,

$E = E°_{Hg^{2+},Hg} - (2.303\ RT/2F) \log [K_{f(Hg-EDTA)}/K_{f(M-EDTA)} - (2.303\ RT/2F)$

$\log (a_{M-EDTA}/a_{Hg-EDTA}) - (2.303\ RT/2F) \log (1/a_{M^{n+}})$

16. From the appendix, the potential of the calomel electrode in 1 $\underline{M}$ KCl is +0.282 V.

$E_{vs\ S.C.E.} = E_{ind} - E_{S.C.E.}$

$-0.465\ V = E_{ind} - 0.242\ V$

$E_{ind} = -0.223\ V$

$E_{vs\ N.C.E.} = E_{ind} - E_{N.C.E.} = -0.223\ V - 0.282\ V = -0.505\ V$

That is, since the N.C.E. is 0.040 V more positive, the indicator electrode will be 0.040 V more negative relative to this electrode.

17. (a) About 0.02 pH unit. It is limited by the accuracy to which the pH of the standardizing solution is known, which is limited in turn by the certainty to which the activity of a single ion can be calculated; that ion is Cl^- in the reference AgCl/Ag electrode used in standardizing measurements and by the residual liquid junction potential.

$E = 0.059\ \Delta pH = 0.059(0.02) = 0.0012\ V = 1.2\ mV$

For n = 1, the error is 4% per mV. Therefore, the error in hydrogen ion activity is 4.8%.

(b) About 0.002 pH unit.

$\Delta E = (0.059)(0.002) = 0.00012\ V = 0.12\ mV$

(4%/mV)(0.12 mV) = 0.48% variation in hydrogen ion activity

18. $E = k - 0.05916\ pH$

$0.395 = k - (0.05916)(7.00)$

$k = 0.809$

(a) $pH = (k - E)/(0.05916) = (0.809 - 0.467)/(0.05916) = 5.78$

(b) $pH = (0.809 - 0.209)/(0.05916) = 10.14$

(c) $pH = (0.809 - 0.080)/(0.05916) = 12.32$

(d) $pH = (0.809 + 0.013)/(0.05916) = 13.89$

19. (a) pH = 3.00

$E_{cell} = E_{H^+,H_2} - E_{S.C.E.} = -0.05916 \log (1/a_{H^+}) - 0.242$

$= -0.05916\ pH - 0.242 = -0.05916\ (3.00) - 0.242 = -0.419\ V$

(i.e., the potential of the indicating electrode is -0.419 V relative to the reference electrode; or $E = 0.419\ V$ with the reference electrode the more positive.)

(b) $H^+ = \sqrt{K_a[HOAc]} = \sqrt{(1.75 \times 10^{-5})(1.00 \times 10^{-3})} = 1.32 \times 10^{-4}$

$pH = 3.88$

$E_{cell} = (0.05916)(3.88) - 0.242 = -0.472\ V$

($\Delta E = 0.472\ V$ with the reference electrode the more positive)

(c) $pH = pK_a + \log([OAc^-]/[HOAc]) = 4.76 + \log 1.00 = 4.76$

$E_{cell} = 0.05916\ (4.76) - 0.242 = -0.524\ V$

($\Delta E = 0.524\ V$ with the reference electrode the more positive)

20. $E_{cell} = E_{ind} - E_{ref} = E_{ind} - 0.242$

In this case, the indicating electrode potential is less positive than that of the S.C.E.

$-0.205 = 0.699 - (0.05916/2) \log [a_{HQ}/(a_Q a_{H^+}^2)] - 0.242$

Since the ratio a_{HQ}/a_Q is unity,

$-0.662 = -(0.05916/2) \log (1/a_{H^+}^2) = 0.05916 \log a_H^+ = -0.05916\ pH$

$pH = 11.2$

21. In a mixture of A^+ and B^+, the electrode potential is given by:

$E_{AB} = k_A + S \log (a_{A^+} + K_{AB} a_{B^+})$ (1)

In a solution containing only B^+, it is given by:

$E_B = k_B + S \log a_{B^+}$ (2)

Also, in a solution containing only B^+, (1) reduces to:

$E_B = k_A + S \log K_{AB} a_{B^+} = k_A + S \log K_{AB} + S \log a_{B^+}$ (3)

Hence, (2) and (3) are equal:

$$k_B + S \log a_{B^+} = k_A + S \log K_{AB} + S \log a_{B^+}; \quad \log K_{AB} = (k_B - k_A)/S \qquad (4)$$

22.. The ionic strength of the KCl standard is 0.0050. From Equation 6.20 in Chapter 6,

$$-\log f_i = 0.51\ (1)^2\ \sqrt{0.0050}/(1 + \sqrt{0.0050})$$

$$f_i = 0.93$$

$$a_{K^+} = 0.0050\ \underline{M} \times 0.93 = 0.0046\ \underline{M}$$

Standard:

$$-18.3\ \text{mV} = k + 59.2 \log (0.0046)$$

$$k = +120.1\ \text{mV}$$

Sample:

$$E = k + 59.2 \log (a_{K^+} + K_{KCs}a_{Cs^+})$$

$$20.9 = +120.1 + 59.2 \log (a_{K^+} + 1 \times 0.0060)$$

$$-1.68 = \log (a_{K^+} + 0.006)$$

$$a_{K^+} = 0.015\ \underline{M}$$

23. Calculate S:

$$-108.6 = k - S \log (0.0050) \qquad (1)$$

$$-125.2 = k - S \log (0.0100) \qquad (2)$$

---

(1)-(2): $16.6 = -S \log (0.0050) + S \log (0.0100) = S \log (0.0100)/(0.0050)$

$$S = 55.1$$

Calculate k. Use either (1) or (2):

$$-125.2\ \text{mV} = k - 55.1 \log (0.0100)$$

$$k = -235.4\ \text{mV}$$

For the sample:

$-119.6 = -235.4 - 55.1 \log [NO_3^-]$

$[NO_3^-] \doteq 0.0079$ $\underline{M}$

24. For ClO_4^-:

$-27.2 = k_{ClO_4^-} - 59.2 \log (0.00100)$; $k_{ClO_4^-} = -204.8$ mV

For I^-:

$+32.8 = k_{I^-} - 59.2 \log (0.0100) = k_{ClO_4^-} - 59.2 \log [K_{ClO_4^-,I^-}(0.0100)]$

$k_{I^-} = -85.6$ mV

$\log K_{ClO_4^-,I^-} = (k_{ClO_4^-} - k_{I^-})/59.2 = (-204.8 + 85.6)/59.2 = -2.01$

$K_{ClO_4^-,I^-} = 0.0097$

For sample:

$-15.5 = -204.8 - 59.2 \log ([ClO_4^-] + 9.7 \times 10^{-3}(0.015))$

$-3.20 = \log ([ClO_4^-] + 1.5 \times 10^{-4})$

$[ClO_4^-] = 6.3 \times 10^{-4} - 1.5 \times 10^{-4} = 4.8 \times 10^{-4}$ $\underline{M}$

25. (a) $k_{Na} = E_{Na} - 59.2 \log a_{Na^+} = 113.0 - 59.2 \log 1.00 \times 10^{-1} = 172.2$ mV

$k_K = 67.0 - 59.2 \log 1.00 \times 10^{-1} = 126.2$ mV

$\log K_{NaK} = (k_K - k_{Na})/59.2 = (126.2 - 172.2)/59.2 = -0.777$

$K_{NaK} = 0.167$

(b) $E_{NaK} = k_{Na} + 59.2 \log (a_{Na^+} + K_{NaK}a_{K^+})$

$= 172.2 + 59.2 \log (1.00 \times 10^{-3} + 0.167 \times 1.00 \times 10^{-2}) = 19.8$ mV

26. $+237.8 = k + 56.1 \log [2.00 \times 10^{-4} + K_{AB} (1.00 \times 10^{-3})]$ (1)

$+253.6 = k + 56.1 \log [4.00 \times 10^{-4} + K_{AB} (1.00 \times 10^{-3})]$ (2)

(2)-(1):

$20.4 = 56.1 \log [4.00 \times 10^{-4} + K_{AB}(1.00\times10^{-3})]/[2.00\times10^{-4}+K_{AB}(1.00\times10^{-3})]$

$$1.91\ [2.00 \times 10^{-4} + K_{AB}\ (1.00 \times 10^{-3})] = 4.00 \times 10^{-4} + K_{AB}\ (1.00 \times 10^{-3})$$

$$K_{AB} = 0.020$$

27. $-\log K_{NaK} = E_{Na} - E_K/S = E/S = 175.5/58.1 = 3.02$

$K_{NaK} = 10^{-3.02} = 10^{-4} \times 10^{.98} = 9.5 \times 10^{-4}$

28. *From Equation 13.52, the electrode responds equally to the two ions when the deviation from the experimental curve is (0.301)(57.8)/1 = 17.4 mV.*

From Equation 13.50,

$K_{KNa} = a_K/a_{Na} = 0.015\ mM/140\ mM = 1.0_7 \times 10^4$

1. *Starch indicator for iodometric and iodimetric titrations.*

   Self indicator for highly colored titrants, e.g., $KMnO_4$

   Redox indicators for other titrations. The indicator is itself a weak oxidizing or reducing agent whose oxidized and reduced forms are different colors. E^o_{ind} should be near the end point potential.

2. *In iodimetry, a reducing agent is titrated with a standard solution of I_2 to a blue starch endpoint. In iodometry, an oxidizing agent is reacted with an excess of I^- to form an equivalent amount of I_2, which is then titrated with a standard solution of $Na_2S_2O_3$ to a starch endpoint (disappearance of blue I_2-starch color).*

3. *Many oxidizing agents that react with iodide consume protons in the reaction, and the equilibrium lies far to the right only in acid solution. Conversely, most reducing agents that react with I_2 liberate protons, and the solution must be near neutrality for the equilibrium to lie far to the right. Also, in iodimetry, the pH is kept near neutrality to minimize acid hydrolysis of the starch and air oxidation of the iodide produced.*

4. *No. It is slightly beyond the equivalence point, because an excess of permanganate must be added to impart a pink color to the solution. A blank titration can be used to correct for the excess.*

5. *It contains $MnSO_4$ and H_3PO_4. The Mn^{2+} is added to decrease the potential of the MnO_4^-/Mn^{2+} couple, so that permanganate will not oxidize Cl^-. The H_3PO_4 is added to complex the Fe^{3+} and decrease the potential of the Fe^{3+}/Fe^{2+} couple and hence sharpen the endpoint. The iron(III) complex is also colorless, which makes the endpoint easier to see.*

6. *We use the procedure in which individual half-reactions are balanced without knowledge of the oxidation states. The O's are balanced with H_2O, the H's with H^+ (followed by neutralization with OH^- if alkaline), and finally, the charges with electrons.*

(a) (1) $IO_3^- = 1/2\ I_2$

(2) $IO_3^- = 1/2\ I_2 + 3H_2O$

(3) $IO_3^- + 6H^+ = 1/2\ I_2 + 3H_2O$

(4) $IO_3^- + 6H^+ + 5e^- = 1/2\ I_2 + 3H_2O$

(1) $I^- = 1/2\ I_2$

(2) $I^- = 1/2\ I_2 + e^-$

$IO_3^- + 6H^+ + 5e^- = 1/2\ I_2 + 3H_2O$

$5(I^- = 1/2\ I_2 + e^-)$

$IO_3^- + 5I^- + 6H^+ = 3\ I_2 + 3H_2O$

(b) (1) $Se_2Cl_2 = 2H_2SeO_3 + 2Cl^-$

(2) $Se_2Cl_2 + 6H_2O = 2H_2SeO_3 + 2Cl^-$

(3) $Se_2Cl_2 + 6H_2O = 2H_2SeO_3 + 2Cl^- + 8H^+$

(4) $Se_2Cl_2 + 6H_2O = 2H_2SeO_3 + 2Cl^- + 8H^+ + 6e^-$

(1) $Se_2Cl_2 = 2Se + 2Cl^-$

(2) $Se_2Cl_2 + 2e^- = 2Se + 2Cl^-$

$Se_2Cl_2 + 6H_2O = 2H_2SeO_3 + 2Cl^- + 8H^+ + 6e^-$

$3(Se_2Cl_2 + 2e^- = 2Se + 2Cl^-)$

$4Se_2Cl_2 + 6H_2O = 2H_2SeO_3 + 6Se + 8Cl^- + 8H^+$

Divide by 2: $2Se_2Cl_2 + 3H_2O = H_2SeO_3 + 3Se + 4Cl^- + 4H^+$

(c) (1) $H_3PO_3 = H_3PO_4$

(2) $H_3PO_3 + H_2O = H_3PO_4$

(3) $H_3PO_3 + H_2O = H_3PO_4 + 2H^+$

(4) $H_3PO_3 + H_2O = H_3PO_4 + 2H^+ + 2e^-$

(1) $2HgCl_2 = Hg_2Cl_2 + 2Cl^-$

(2) $2HgCl_2 + 2e^- = Hg_2Cl_2 + 2Cl^-$

$H_3PO_3 + H_2O = H_3PO_4 + 2H^+ + 2e^-$

$2HgCl_2 + 2e^- = Hg_2Cl_2 + 2Cl^-$

$H_3PO_3 + 2HgCl_2 + H_2O = H_3PO_4 + Hg_2Cl_2 + 2H^+ + 2Cl^-$

7. (a) (1) $MnO_4^{2-} = MnO_2$

(2) $MnO_4^{2-} = MnO_2 + 2H_2O$

(3) $MnO_4^{2-} + 4H^+ = MnO_2 + 2H_2O$

$MnO_4^{2-} + 2H_2O = MnO_2 + 4OH^-$

(4) $MnO_4^{2-} + 2H_2O + 2e^- = MnO_2 + 4OH^-$

(1) $MnO_4^{2-} = MnO_4^-$

(2) $MnO_4^{2-} = MnO_4^- + e^-$

$MnO_4^{2-} + 2H_2O + 2e^- = MnO_2 + 4OH^-$

$2(MnO_4^{2-} = MnO_4^- + e^-)$

$3MnO_4^{2-} + 2H_2O = MnO_2 + 2MnO_4^- + 4OH^-$

(b) (1) $MnO_4^- = Mn^{2+}$

(2) $MnO_4^- = Mn^{2+} + 4H_2O$

(3) $MnO_4^- + 8H^+ = Mn^{2+} + 4H_2O$

(4) $MnO_4^- + 8H^+ + 5e^- = Mn^{2+} + 4H_2O$

(1) $H_2S = S$

(2) $H_2S = S + 2H^+$

(3) $H_2S = S + 2H^+ + 2e^-$

$2(MnO_4^- + 8H^+ + 5e^- = Mn^{2+} + 4H_2O)$

$5(H_2S = S + 2H^+ + 2e^-)$

$2MnO_4^- + 5H_2S + 6H^+ = 2Mn^{2+} + 5S + 8H_2O$

(c) (1) $2SbH_3 = H_4Sb_2O_7$

(2) $2SbH_3 + 7H_2O = H_4Sb_2O_7$

(3) $2SbH_3 + 7H_2O = H_4Sb_2O_7 + 16H^+$

(4) $2SbH_3 + 7H_2O = H_4Sb_2O_7 + 16H^+ + 16e^-$

(1) $Cl_2O = 2Cl^-$

(2) $Cl_2O = 2Cl^- + H_2O$

(3) $Cl_2O + 2H^+ = 2Cl^- + H_2O$

(4) $Cl_2O + 2H^+ + 4e^- = 2Cl^- + H_2O$

$2SbH_3 + 7H_2O = H_4Sb_2O_7 + 16H^+ + 16e^-$

$4(Cl_2O + 2H^+ + 4e^- = 2Cl^- + H_2O)$

$2SbH_3 + 4Cl_2O + 3H_2O = H_4Sb_2O_7 + 8Cl^- + 8H^+$

(d) (1) $FeS = Fe^{3+} + S$

(2) $FeS = Fe^{3+} + S + 3e^-$

(1) $NO_3^- = NO_2$

(2) $NO_3^- = NO_2 + H_2O$

(3) $NO_3^- + 2H^+ = NO_2 + H_2O$

(4) $NO_3^- + 2H^+ + e^- = NO_2 + H_2O$

$FeS = Fe^{3+} + S + 3e^-$

$3(NO_3^- + 2H^+ + e^- = NO_2 + H_2O)$

$FeS + 3NO_3^- + 6H^+ = Fe^{3+} + S + 3NO_2 + 3H_2O$

(e) (1) $Al = AlO_2^-$

(2) $Al + 2H_2O = AlO_2^-$

(3) $Al + 2H_2O = AlO_2^- + 4H^+$

$Al + 4OH^- = AlO_2^- + 2H_2O$

(4) $Al + 4OH^- = AlO_2^- + 2H_2O + 3e^-$

(1) $NO_3^- = NH_3$

(2) $NO_3^- = NH_3 + 3H_2O$

(3) $NO_3^- + 9H^+ = NH_3 + 3H_2O$

$NO_3^- + 6H_2O = NH_3 + 9OH^-$

(4) $NO_3^- + 6H_2O + 8e^- = NH_3 + 9OH^-$

$8(Al + 4OH^- = AlO_2^- + 2H_2O + 3e^-)$

$3(NO_3^- + 6H_2O + 8e^- = NH_3 + 9OH^-)$

$8Al + 3NO_3^- + 5OH^- + 2H_2O = 8AlO_2^- + 3NH_3$

(f) (1) $FeAsS = Fe^{3+} + AsO_4^{3-} + SO_4^{2-}$

(2) $FeAsS + 8H_2O = Fe^{3+} + AsO_4^{3-} + SO_4^{2-}$

(3) $FeAsS + 8H_2O = Fe^{3+} + AsO_4^{3-} + SO_4^{2-} + 16H^+$

(4) $FeAsS + 8H_2O = Fe^{3+} + AsO_4^{3-} + SO_4^{2-} + 16H^+ + 14e^-$

(1) $ClO_2 = Cl^-$

(2) $ClO_2 = Cl^- + 2H_2O$

(3) $ClO_2 + 4H^+ = Cl^- + 2H_2O$

(4) $ClO_2 + 4H^+ + 5e^- = Cl^- + 2H_2O$

$5(FeAsS + 8H_2O = Fe^{3+} + AsO_4^{3-} + SO_4^{2-} + 16H^+ + 14e^-)$

$14(ClO_2 + 4H^+ + 5e^- = Cl^- + 2H_2O)$

$5FeAsS + 14ClO_2 + 12H_2O = 5Fe^{3+} + 5AsO_4^{3-} + 5SO_4^{2-} + 14Cl^- + 24H^+$

(g) (1) $K_2NaCo(NO_2)_6 = 2K^+ + Na^+ + Co^{3+} + 6NO_3^-$

(2) $K_2NaCo(NO_2)_6 + 6H_2O = 2K^+ + Na^+ + Co^{3+} + 6NO_3^-$

(3) $K_2NaCo(NO_2)_6 + 6H_2O = 2K^+ + Na^+ + Co^{3+} + 6NO_3^- + 12H^+$

(4) $K_2NaCo(NO_2)_6 + 6H_2O = 2K^+ + Na^+ + Co^{3+} + 6NO_3^- + 12H^+ + 12e^-$

$5(K_2NaCo(NO_2)_6 + 6H_2O = 2K^+ + Na^+ + Co^{3+} + 6NO_3^- + 12H^+ + 12e^-)$

$12(MnO_4^- + 8H^+ + 5e^- = Mn^{2+} + 4H_2O)$

$5K_2NaCo(NO_2)_6 + 12MnO_4^- + 36H^+ = 10K^+ + 5Na^+ + 5Co^{3+} + 30NO_3^- + 12Mn^{2+} + 18H_2O$

8. $Tl^+ + 2Co^{3+} = Tl^{3+} + 2Co^{2+}$

Equivalent amounts of Tl^+ and Co^{3+} are added. There are no polynuclear species nor H^+

$\therefore$ $E = (2 \times 0.77 + 1 \times 1.1842)/(2 + 1) = 1.12_7$ V

9. $6Fe^{2+} + Cr_2O_7^{2-} + 14H^+ = 6Fe^{3+} + 2Cr^{3+} + 7H_2O$

Before the endpoint, the half cell

$Fe^{3+} + e^- = Fe^{2+}$

is used for the potential calculation, since we know the concentration of each species. After the endpoint, the half cell

$Cr_2O_7^{2-} + 14H^+ + 6e^- = 2Cr^{3+} + 7H_2O$

is used.

We start with 5.00 mmol Fe^{2+} and 50.0 mmol H^+

10 mL: mmol Fe^{3+} = 1.00

mmol Fe^{2+} = 4.00

mmol Cr^{3+} = 0.334

mmol H^+ = 47.7

$E = 0.771 - 0.059 \log (4.00/1.00) = 0.735$ V

25 mL: mmol Fe^{3+} = 2.50

mmol Fe^{2+} = 2.50

mmol Cr^{3+} = 0.835

mmol H^+ = 44.2

$E = 0.771 - 0.059 \log (2.50/2.50) = 0.771$ V

50 mL: mmol Fe^{3+} = 5.00 - x ≈ 5.00

mmol Fe^{2+} = x

mmol Cr^{3+} = 1.67 - 1/3 x = 1.67; M = 0.0167 mmol/mL

mmol $Cr_2O_7^{2-}$ = 1/6 x; M = (1/6 x/100) mmol/mL

mmol H^+ = 38.3 + 14/6 x ≈ 38.3; M = 0.383 mmol/mL

Calculate K_{eq}:

$0.771 - (0.059/6) \log ([Fe^{2+}]^6/[Fe^{3+}]^6)$

$= 1.33 - (0.059/6) \log ([Cr^{3+}]^2/[Cr_2O_7^{2-}][H^+]^{14})$

$-\log K_{eq} = -\log ([Fe^{3+}]^6[Cr^{3+}]^2)/([Fe^{2+}]^6[Cr_2O_7^{2-}][H^+]^{14})$

$= (0.771 - 1.33)/(0.059/6)$

$K_{eq} = 7 \times 10^{56}$

Use millimoles Fe^{3+} and Fe^{2+} since these volumes cancel, but use molarities of others:

$[(5.00)^6(0.0167)^2]/[(x)^6(x/600)(0.383)^{14}] = 7 \times 10^{56}$

$x = 1._3 \times 10^{-7}$ mmol Fe^{2+}

$E = 0.771 - 0.059 \log (1._3 \times 10^{-7}/5.00) = 1.218$ V

60 mL: mmol Fe^{3+} = 5.00 - x ≈ 5.00

mmol Fe^{2+} = x

mmol Cr^{3+} = 1.67 - 1/3 x ≈ 1.67; M = 0.0152 mmol/mL

mmol $Cr_2O_7^{2-}$ = 0.167 + 1/6 x ≈ 0.167; M = 0.00152 mmol/mL

mmol H^+ = 38.3 - 14/6 x ≈ 38.3; M = 0.348 mmol/mL

$E = 1.33 - (0.059/6) \log [(0.0152)^2/(0.00152)(0.348)^{14}] = 1.28$ V

10. 10.0 mL: mmol MnO_4^- added = 0.0200 M x 10.0 mL = 0.200 mmol

mmol Fe^{2+} reacted = 5 x 0.200 = 1.00 mmol = mmol Fe^{3+} formed

mmol Fe^{2+} left = 0.100 M x 100 mL - 1.00 mmol = 9.00 mmol Fe^{2+}

$E = 0.771 - 0.059 \log (9.00)/(1.00) = 0.715$ V

50.0 mL: One half the Fe^{2+} is converted to Fe^{3+} (5.00 mmol each)

E = 0.771 V

100 mL: See Example 14.6.

E = 1.36 V

200 mL: mmol Mn^{2+} = 2.00 - 1/5 x ≈ 2.00 mmol

mmol MnO_4^- = 0.0200 M x 100 mL + 1/5 x ≈ 2.00 mmol

$[H^+]$ = 84 mmol/300 mL = 0.28 M

(We would have to calculate x as at 100 mL to use the Fe equation.)

$E = 1.51 - (0.059)/(5) \log (2.00)/(2.00)(0.28)^8 = 1.46$ V

11. This is a symmetrical reaction, so:

$E_{e.p.} = (n_1E°_1+n_2E°_2)/(n_1+n_2) = [(1)(0.771)+(2)(0.154)]/(1+2) = 0.360$ V

12. $2Fe^{3+} + Sn^{2+} = 2Fe^{2+} + Sn^{4+}$

2x x 2c-2x c-x

Neglecting x compared to c:

$E = E°_{Fe^{3+},Fe^{2+}} - (0.059/n_{Fe})\ \log\ ([Fe^{2+}]/[Fe^{3+}]) = E°_{Fe^{3+},Fe^{2+}}$

$- (0.059/1)\ \log\ (2c/2x)$ (1)

$E = E°_{Sn^{4+},Sn^{2+}} - (0.059/n_{Sn})\ \log\ ([Sn^{2+}]/[Sn^{4+}]) = E°_{Sn^{4+},Sn^{2+}}$

$- (0.059/2)\ \log\ (x/c)$ (2)

From (1):

$(1)\ \ E = (1)\ E°_{Fe^{3+},Fe^{2+}} - 0.059\ \log\ (c/x)$ (3)

From (2):

$(2)\ E = (2)\ E°_{Sn^{4+},Sn^{2+}} - 0.059\ \log\ (x/c)$ (4)

Adding (3) and (4):

$(1)E + (2)E = (1)E°_{Fe^{3+},Fe^{2+}} + (2)E°_{Sn^{4+},Sn^{2+}} - 0.059\ \log\ [(c/x)(x/c)]$

$E = [(1)E°_{Fe^{3+},Fe^{2+}} + (2)E°_{Sn^{4+},Sn^{2+}}]/[(1) + (2)]$

13. $5Fe^{2+} + MnO_4^- + 8H^+ = 5Fe^{3+} + Mn^{2+} + 4H_2O$

5x x 5c−5x c−x

Neglecting x compared to c:

$E = E°_{Fe^{3+},Fe^{2+}} - (0.059/n_{Fe})\ \log\ ([Fe^{2+}]/[Fe^{3+}]);$

$n_{Fe}E = n_{Fe}E°_{Fe^{3+},Fe^{2+}} - 0.059\ \log\ (5x/5c)$ (1)

$E = E°_{MnO_4^-,Mn^{2+}} - (0.059/n_{Mn})\ \log\ ([Mn^{2+}]/[MnO_4^-][H^+]^8;$

$n_{Mn}E = n_{Mn}E°_{MnO_4^-,Mn^{2+}} = 0.059\ \log\ (c/x) + 0.059\ \log\ [H^+]^8$ (2)

Adding (1) and (2):

$n_{Fe}E + n_{Mn}E = n_{Fe}E°_{Fe^{3+},Fe^{2+}} + n_{Mn}E°_{MnO_4^-,Mn^{2+}}$

$- 0.059\ \log\ [(5x/5c)(c/x)] + (8)(0.059)\ \log\ [H^+]$

$E = (n_{Fe}E°_{Fe^{2+},Fe^{3+}} + n_{Mn}E°_{MnO_4^-,Mn^{2+}})/(n_{Fe}+n_{Mn}) - [(8)(0.059)/(n_{Fe}+n_{Mn})]$ pH

$E_{e.p.} = [(1)(0.771)+(5)(1.51)]/(1+5) - [(8)(0.059)/(1+5)] \times 0.38 = 1.35_7$ V

This compares with 1.35_9 V obtained in the example calculating the equilibrium concentrations.

14. (a) Before reaction:

$E_{cell} = -0.403 - (0.059/2)\log(1/[Cd^{2+}]) + 0.763 + (0.059/2)\log(1/[Zn^{2+}])$

$= -0.403 - (0.059/2)\log(1/0.0100) + 0.763 + (0.059/2)\log(1/0.250) = 0.319$ V

After reaction, $E_{cell} = 0$. The reaction is:

$Zn + Cd^{2+} = Zn^{2+} + Cd$

$x \quad 0.260-x$

≈ 0.260 M

Assuming excess metallic zinc, the 0.0100 M Cd^{2+} has produced 0.0100 M Zn^{2+} and the total $[Zn^{2+}]$ is 0.260 M (x is the equilibrium concentration of Cd^{2+}). Since $E_{Cd} = E_{Zn}$,

$E_{Cd} = E_{Zn} = -0.763 - (0.059/2)\log(1/0.260) = -0.780$ V

Equilibrium constant:

$-0.403 - (0.059/2)\log(1/[Cd^{2+}]) = -0.763 - (0.059/2)\log(1/[Zn^{2+}])$

$-\log K_{eq} = -\log([Zn^{2+}]/[Cd^{2+}]) = (-0.360)/(0.059/2)$

$K_{eq} = 1.6 \times 10^{12}$

(b) Before reaction:

$E_{cell} = 0.5355 - (0.059/2)\log [I^-]^3/[I_3^-] + 0.126 + (0.059/2)\log(1/[Pb^{2+}])$

$= 0.5355 - (0.059/2)\ \log[(1.00)^3/(0.100)] + 0.126 + (0.059/2)\ \log(1/0.0100)$

$= 0.691\ V$

After reaction:

$Pb + I_3^- = Pb^{2+} \quad + 3I^-$

$x \quad 0.110-x \quad 1.30-3x$

$\approx 0.110\ \underline{M} \quad \approx 1.30\ \underline{M}$

$E_{Pb} = E_I = -0.126 - (0.059/2)\ \log\ (1/0.110) = -0.154\ V$

Equilibrium constant:

$0.5355 - (0.059/2)\ \log\ ([I^-]^3/[I_3^-]) = -0.126 - (0.059/2)\ \log\ (1/[Pb^{2+}])$

$-\log\ K_{eq} = -\log\ ([Pb^{2+}][I^-]^3/[I_3^-]) = (-0.126 - 0.5355)/(0.059/2)$

$K_{eq} = 2.6 \times 10^{22}$

15. mmol $S_2O_3^{2-} = 6 \times$ mmol $K_2Cr_2O_7$ (Equations 14.9 and 14.10)

$\therefore\ \underline{M} \times 1\ mL = [(0.0490\ mg)/(294\ mg/mmol)] \times 6$

$S_2O_3^{2-} = 0.00100\ \underline{M}$

$SeO_3^{2-} + 4I^- + 6H^+ = Se + 2I_2 + 3H_2O$

$\therefore$ mmol $SeO_3^{2-} = 1/2$ mmol $I_2 = 1/4$ mmol $S_2O_3^{2-}$

mg Se = $0.00100\ \underline{M} \times 4.5\ mL \times 1/4 \times 79.0$ mg/mmol = 0.089 mg = 89 μg

89 μg/10.0 g = 8.9 μg/g = 8.9 ppm Se

16. $5H_2C_2O_4 + 2MnO_4^- + 6H^+ = 10\ CO_2 + 2Mn^{2+} + 8H_2O$

mmol Ca^{2+} = mmol $C_2O_4^{2-}$ = 5/2 mmol MnO_4^-

$\therefore$ mmol Ca^{2+} = 5/2 x 0.00100 M x 4.94 mL = 0.0124 mmol

meq Ca^{2+} = 2 x mmol Ca^{2+} = 2 x 0.0124 = 0.0248 meq

$[Ca^{2+}]$ = 0.0248 meq/5.00 mL = 4.96 meq/L

17. The first titration measures only the As(III) (Na_2HAsO_3).

$H_2AsO_3^- + I_2 + H_2O = HAsO_4^{2-} + 2I^- + 3H^+$

mmol As(III) = mmol I_2

% Na_2HAsO_3 = [(0.150 M x 11.3 mL x 170 mg/mmol)/(2500 mg)] x 100% = 11.5%

The second titration measures total arsenic; the As(III) that was converted to As(V) in the first titration, plus the As(V) initially present.

$H_3AsO_4 + 2I^- + 2H^+ = H_3AsO_3 + I_2 + H_2O$

mmol As = mmol I_2 = 1/2 mmol $Na_2S_2O_3$

mmol total As = 1/2 x 0.120 M x 41.2 mL = 2.47 mmol

mmol As(III) = 0.150 M x 11.3 mL = 1.70 mmol

mmol As(V) = 2.47 - 1.70 = 0.77 mmol

mmol As_2O_5 = 1/2 mmol As(V)

$\therefore$ % As_2O_5 = [(0.77 mmol x 1/2 x 230 mg/mmol)/(2500 mg)] x 100% = 3.5_4%

18. M_{KMnO_4} = [125 mg Fe x 1/5 (mmol MnO_4^-/mmol Fe)]/(55.8 mg/mmol x 1.00 mL $KMnO_4$)

= 0.448 M

M_{tetrox} = [0.448 M_{KMnO_4} (mmol/mL) x 0.175 mL_{KMnO_4}

x 5/4 (mmol tetrox/mmol $KMnO_4$)]/(1.00 mL_{tetrox}) = 0.0980 M

$$0.200\ \underline{M}_{NaOH}\ (mmol/mL) \times mL_{NaOH} = 0.0980\ \underline{M}_{tetrox}\ (mmol/mL)$$
$$\times\ 1.00\ mL_{tetrox} \times 3\ (mmol\ NaOH/mmol\ tetrox)$$
$$mL_{NaOH} = 1.47\ mL$$

19. The concentrations of standards in the diluted samples are 1.2×10^{-3} $\underline{M}$ and 2.4×10^{-3} $\underline{M}$, respectively. Let x represent the concentration of sulfide in the diluted sample.

$$-216.4 = k - 29.6 \log x \qquad (1)$$
$$-224.0 = k - 29.6 \log (x + 1.2 \times 10^{-3}) \qquad (2)$$

(1)-(2):

$$7.6 = 29.6 \log [(x + 1.2 \times 10^{-3})/x]$$
$$(x + 1.2 \times 10^{-3})/x = 1.81$$
$$x = 1.5 \times 10^{-3}\ \underline{M}$$

The sample was diluted 2.5:1. Therefore, the sample concentration is:

$$(1.5 \times 10^{-3}) \times 2.5 = 3.8 \times 10^{-3}\ \underline{M}\ \text{sulfide}$$

20. From K_a,

$$[H^+] = K_a[HA]/[A^-]$$

During the titration,

$$[HA] = (M_{HA}V_{HA} - M_BV_B)/(V_{HA} + V_B)$$
$$[A^-] = M_BV_B/(V_{HA} + V_B)$$

So

$$[H^+] = K_a(M_{HA}V_{HA} - M_BV_B)/M_BV_B$$

At the equivalence point

$M_{HA}V_{HA} = M_BV_{eq\,pt}$

So

$[H^+] = K_a(M_BV_{eq\,pt} - M_BV_B)/M_BV_B$

or

$V_B[H^+] = K_a(V_{eq\,pt} - V_B) = V_B\,10^{-pH}$ (since $[H^+] = 10^{-pH}$)

At the equivalence point, $V_B = V_{eq\,pt}$, *and so the equivalence point corresponds to the zero intercept in the y-axis in a plot of* V_B10^{-pH} *vs.* V_B. *The slope is* $-K_a$.

21. *See CD for spreadsheet calculation of titration curve and the plot.*

CHAPTER 15

1. *Back emf = voltage exerted by a galvanic cell which opposes the external applied voltage required for electrolysis to occur*

   Overpotential = the electrode potential in excess of the reversible Nernst potential required for electrolysis to occur

   IR drop = the voltage drop in an electrolysis cell due to the resistance of the cell and is equal to the product of the cell current and the cell resistance. An increased voltage, equal to the iR drop, is required to maintain the electrolysis.

2. *Half-wave potential = the potential of a voltammetric wave at which the current is one-half the limiting current. Depolarizer = a substance that is reduced or oxidized at a microelectrode*

   D.M.E. = dropping mercury electrode. This is the microelectrode used for polarography.

   Residual current = the background voltammetric current, caused by impurities and the charging current. This limits the sensitivity of these techniques.

   Voltammetry = all current-voltage methods using microelectrodes

3. *The supporting electrolyte minimizes the migration current and the iR drop.*

4. *Electrolytically reduce the* Fe^{3+} *to* Fe^{2+} *at 0 V vs S.C.E at a large cathode before running the polarogram, using either a platinum gauze or a mercury pool cathode. The* Fe^{2+} *will not be reduced at -0.4 V and will not interfere with the polarographic lead wave.*

5. *The complexed metal is stabilized against reduction, and a more negative potential will be required to reduce it, i.e., the half-wave potential is shifted to a more negative potential.*

6. *A chemically modified electrode consists of an electrochemical transducer and a chemically selective (or catalytic) layer.*

7. *The electrocatalyst reduces the needed applied potential to generate an amperometric current from an electrochemically irreversible analyte. It does so by reacting with the analyte to be converted to a form which is electrolyzed at the applied potential.*

8. The dimensions of an ultramicroelectrode are smaller than the diffusion layer thickness, making the mass transport independent of flow and enhancing the signal-to-noise ratio.

9. Concentration of lead added = C; x = concentration in unknown solution.

$C \times 11.0\ mL = 1.00 \times 10^{-3}\ \underline{M} \times 1.00\ mL$

$C = 9.1 \times 10^{-5}\ \underline{M}$

Current increase caused by added lead = (12.2 - 5.6) μA = 6.6 μA

$(6.6\ \mu A)/(9.1 \times 10^{-5}\ \underline{M}) = (5.6\ \mu A)/(x)$

$x = 7.7 \times 10^{-5}\ \underline{M}$ lead

10. Since a current is recorded at zero V, Fe^{3+} is present. Since the wave at -1.5 V exceeds twice the current at zero V, Fe^{2+} is also present. The Fe^{3+} contribution to the wave at -1.5 V is 2 x 12.5 = 25.0 μA. Therefore, the Fe^{2+} contribution is 30.0 - 25.0 = 5.0 μA. The relative concentrations are $[Fe^{3+}]/[Fe^{2+}] = 25.0/5.0 = 5:1$.

1. In the far infared region, quantized rotational energy transitions occur in absorption. In the mid-infared region, these are superimposed on quantized vibrational transitions. In the visible and ultraviolet regions, quantized electronic transitions occur in addition to the rotational and vibrational transitions.

2. Paired nonbonding outershell electrons (n electrons) and pi (π) electrons in double or triple bonds

3. Transitions to a π^* antibonding orbital ($n \rightarrow \pi^*$ and $\pi \rightarrow \pi^*$ transitions). A wavelengths less than 200 nm, $n \rightarrow \sigma^*$ transitions may also occur. $\pi \rightarrow \pi$ transitions are the most intense (ε = 1000 - 100,000 vs. 1000 for $n \rightarrow \pi^*$).

Examples are: $\pi \rightarrow \pi^*$ and $n \rightarrow \pi^*$ - ketones

$n \rightarrow \pi^*$ - ethers, disulfides, alkyl halides, etc. at < 200 nm.

4. There must be a change in the dipole moment of the molecule.

5. Stretching and bending.

6. Absorption in the near-IR region (0.75 - 2.5 mm) is the result of vibrational overtones which are weak and featureless. They are due mainly to C-H, O-H and N-H band stretching and bending motions. NIR is useful for analyzing "neat", i.e., undiluted, samples, and as such, is useful for non-destructive analysis.

7. chromophore - an absorbing group

auxochrome - enhances absorption by a chromophore or shifts its wavelength of absorption

bathochromic shift - λ_{max} shifted to longer wavelength

hypsochromic shift - λ_{max} shifted to shorter wavelength

hyperchromism - absorption intensity increased

hypochromism - absorption intensity decreased

8. (a) CH_2=$CHCO_2H$

(b) CH_3C=C-C=CCH_3

(c)

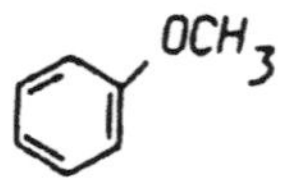

9. (a) Bathochromic shift, increased absorption

(b) bathochromic shift, increased absorption

(c) no shift, but increased absorption

10. They are highly conjugated and in alkaline solution loss of a proton affects the electron distribution and hence the wavelength of absorption.

11. Excitation of the metal ion or of the ligand, or via a charge transfer transition (movement of electrons between the metal ion and ligand).

12. absorption = fraction of light absorbed = $1 - (P/P_0)$

absorbance = $-\log (P/P_0)$ and is proportional to the concentration.

percent transmittance = percent of light transmitted = $(P/P_0) \times 100$

transmittance = fraction of light transmitted = (P/P_0)

13. Absorptivity is the proportionality constant, a, in Beer's law, when the concentration units are expressed in g/L. Molar absorptivity, ε. is the proportionality constant when the concentration is expressed in mol/L and is equal to a x f.w.

14. At an absorption maximum, the average absorptivity of the band of wavelengths passed remains more nearly constant as the concentration is changed. The steepness of an absorption shoulder increases as the concentration increases, with the result that the average absorptivity of the wavelengths passed may change.

15. Ultraviolet region: the solvents listed in Table 16.3.

Visible region: any colorless solvent, including those listed in Table 14.3.

Infared region: carbon tetrachloride or carbon disulfide to cover the region of 2.5-15 μm.

16. An isobestic point is the wavelength at which the absorptivities of two species in equilibrium with each other are equal, i.e., where their absorption spectra overlap. The absorbance at this wavelength remains constant as the equilibrium is shifted, e.g., by varying pH.

17. Deviations from Beer's law can be caused by chemical equilibria in which the equilibria are concentration-dependent. These can usually be minimized by suitable buffering. Other deviations can be caused by instrumental limitations, particularly by the fact that a band of wavelengths is passed by the instrument, rather than monochromatic light. These are apparent deviations, which can be minimized or corrected for. Real deviations from Beer's law occur when the concentration is so high that the index of refraction of the solution is changed, or when the index of refraction of the sample solvent is different from that of the reference solvent.

18.

Region	Source	Detector
Ultraviolet	H_2 or D_2 discharge tube	Phototube or photomultiplier tube
Visible	Tungsten bulb	Phototube or photomultiplier tube
Infrared	Nernst glower or Globar	Thermocouple, balometer, or thermister

19. The prism and the diffraction grating. The prism has nonlinear dispersion and is most useful for dispersion of the shorter wavelengths, e.g., in the uv region. Gratings have linear dispersion and result in equal resolution at all wavelengths, i.e., the infrared region.

20. As the slit width is increased, the resolution is decreased, since a wider range of wavelengths is passed. This results in a narrower concentration range over which Beer's law is obeyed because the light passed is less monochromatic. The spectral slit width is the range of wavelengths passed by the slit, whereas the slit width is the physical dimension (width) of the mechanical slit.

21. In a single beam spectrophotometer, the light from the source passes through the sample and falls on the detector, which directly measures the amount of light absorbed. In a double beam instrument, the light from the source is split, with one path going through the sample, and the other around it. The detector measures each beam alternately and the difference in their intensities is read. This minimizes instability due to fluctuations in the power source, amplifier electronics, etc. It also allows for automatic blank correction when the blank solution is placed in the reference path.

22. Near-IR instruments are reliable because of the more intense radiation sources, high radiation throughput, and sensitive detectors, which results in high signal-to-noise ratios.

23. The radiation is dispersed after passing through the sample, and the dispersed radiation falls on the face of a linear diode array. Each diode acts as an individual detector for the different wavelengths, providing measurement of all wavelengths simultaneously.

24. The radiation source is split into two beams, one reflected by a fixed mirror and the other by a moving mirror at 90° from the other. The two reflected beams combine, resulting in an interference pattern for each wavelength, and pass through the sample. The recorded time domain spectrum is an interferogram. Fourier transformation of this gives a conventional frequency domain spectrum. An interferometer offers advantages of increased throughput and the multiplex advantage of measuring all wavelengths simultaneously. The throughput advantage is particularly valuable in the infrared region where source intensities are weak.

25. The acid form absorbs at 580 nm and the base form absorbs at 450 nm. Referring to Table 16.1, the former would be blue and the latter would be yellow or yellow-green. Filters of the complement colors would be used for each since they must pass only the radiation to be absorbed by each compared, i.e., a yellow filter for the acid form and a blue or violet filter for the base form.

26. Light of high energy (e.g., uv) is absorbed and raises the molecule to a higher electronic energy level. Part of the absorbed energy is lost via collisional processes, whereby the electrons are dropped to the lowest vibrational energy of the first excited state. The electrons return to the ground state from this level, emitting a photon of lower energy and longer wavelength than the absorbed light. Fluorescence is more sensitive than absorption analysis, because the difference between zero and a finite signal is measured, as opposed to the difference between two finite signals in absorption measurements. The sensitivity is then governed by the intensity of the source and the sensitivity and stability of the detector.

27. In dilute solution, when less than about 8% of the light is absorbed ($abc < 0.01$)

28. An intense uv light source at right angles to the detector, a primary filter or monochromator between the source and the sample to pass the excitation wavelength (i.e., reject the fluorescent wavelength), a secondary filter or monochromator between the sample and the detector to pass only the fluorescent wavelength, and a photocell or photomultiplier detector.

29. Measure its degree of fluorescence quenching (an indirect method)

30. $(2500\ \text{Å})(10^{-8}\ \text{cm Å}^{-1})(10^{4}\ \mu\text{m cm}^{-1}) = 0.25\ \mu\text{m}$

$(2500\ \text{Å})(0.1\ \text{nm Å}^{-1}) = 250\ \text{nm}$

31. $\nu = (c/\lambda) = (3.0 \times 10^{10}\ \text{cm s}^{-1})/[(4000\ \text{Å cycle}^{-1})(10^{-8}\ \text{cm Å}^{-1})]$

$= 7.5 \times 10^{14}\ \text{cycles s}^{-1}\ (\text{Hz})$

$\bar{\nu} = (1/\lambda) = 1/[(4000\ \text{Å})(10^{-8}\ \text{cm Å}^{-1})] = 25{,}000\ \text{cm}^{-1}$

32. $(2\ \mu\text{m})(10^{-4}\ \text{cm}\ \mu\text{m}^{-1}) = 2 \times 10^{-4}\ \text{cm}$

$(2 \times 10^{-4}\ \text{cm})(10^{8}\ \text{Å cm}^{-1}) = 20{,}000\ \text{Å}$

$\bar{\nu} = 1/(2 \times 10^{-4}\ \text{cm}) = 5{,}000\ \text{cm}^{-1}$

$(15\ \mu\text{m})(10^{-4}\ \text{cm}\ \mu\text{m}^{-1}) = 1.5 \times 10^{-3}\ \text{cm}$

$(1.5 \times 10^{-3}\ \text{cm})(10^{8}\ \text{Å cm}^{-1}) = 150{,}000\ \text{Å}$

$\bar{\nu} = 1/(1.5 \times 10^{-3}\ \text{cm}) = 670\ \text{cm}^{-1}$

33. Energy of a single photon $= E = h\nu$

$\nu = (3.0 \times 10^{10}\ \text{cm s}^{-1})/[(3000\ \text{Å})(10^{-8}\ \text{cm Å}^{-1})] = 1.0 \times 10^{15}\ \text{s}^{-1}$

$E = (6.6 \times 10^{-34}$ J-s$)(1.0 \times 10^{15}$ s$^{-1}) = 6.6 \times 10^{-19}$ Joule per photon

$(6.6 \times 10^{-19}$ J photon$^{-1})(6.0 \times 10^{23}$ photons einstein$^{-1})$

$= 4.0 \times 10^{5}$ J einstein^{-1}

$(4.0 \times 10^{5}$ J einstein$^{-1})/(4.186$ J cal$^{-1})$

$= 9.5 \times 10^{4}$ cal einstein^{-1}

34. At 20% T, $A = -\log 0.20 = 0.70$

At 80% T, $A = -\log 0.80 = 0.10$

At 0.25 A, $T = 10^{-0.25} = 10^{-1} \times 10^{.75} = 0.56$

At 1.00 A, $T = 10^{-1.00} = 0.10$

35. $A = abc$

$0.80 = a(2 \text{ cm})(0.020 \text{ g/L})$

$a = 20$

36. $\underline{M} = [(15.0\ \mu\text{g/mL})/(280\ \mu\text{g}/\mu\text{mol})] \times 10^{-3}\ \text{mmol}/\mu\text{mol} = 5.36 \times 10^{-5}\ \text{mmol/mL}$

$T = 0.350$

$A = \varepsilon bc = \log (1/T)$

$\varepsilon \times 2 \text{ cm} \times 5.36 \times 10^{-5} \text{ mmol/mL} = \log (1/0.350) = 0.456$

$\varepsilon = 4.25 \times 10^{3}$

37. (a) At 2.00×10^{-5} $\underline{M}$, $T = 1 - 0.325 = 0.685$

$A = \log 1/0.685 = 0.164$

At 6.00×10^{-5} $\underline{M}$, A is tripled:

$A = 3.00 \times 0.164 = 0.492$

(b) $\log 1/T = 0.492$

$1/T = 3.104$

$T = 0.322$; $\%T = 32.2$

% absorbed = 100.0 - 32.2 = 67.8%

38. $\varepsilon = a \times f.w.$

$286\ cm^{-1}g^{-1}\ L \times 180\ g\ mol^{-1} = 5.15 \times 10^4\ cm^{-1}\ mol^{-1}\ L$

39. $A = \varepsilon bc$; f.w. aniline = 93.1

$\varepsilon = a \times f.w. = (134\ cm^{-1}g^{-1}\ L)(93.1\ g\ mol^{-1}) = 1.25 \times 10^4\ cm^{-1}\ mol^{-1}\ L$

$A = (1.25 \times 10^4\ cm^{-1}\ mol^{-1}\ L)(1.00\ cm)(1.00 \times 10^{-4}\ mol\ L^{-}) = 1.25$

40. $A = \varepsilon bc$

$0.687 = 703 \times 1 \times c$

$c = 9.77 \times 10^{-4}\ mol/L$

$9.77 \times 10^{-4}\ mol/L \times 270\ g/mol \times 2\ L = 0.528$ g tolbutamine

41. $A = \varepsilon bc$; *f.w. aniline = 93.1*

$0.425 = 1.25 \times 10^4 \times 1 \times c$

$c = 3.40 \times 10^{-5}\ mol/L$

$3.40 \times 10^{-5}\ mol/L \times 93.1\ g/mol = 3.17 \times 10^{-3}\ g/L$

But the original solution was diluted 250:25 and 100:10 before measurement

∴ total g aniline =

$3.17 \times 10^{-3}\ g/L \times (250/25) \times (100/10) \times 0.500\ L = 0.158\ g$

% aniline $= (0.158\ g/0.200\ g) \times 100\% = 79.0\%$

42. *(a) It is apparent from the calibration data that Beer's law is obeyed. Hence, the concentration in the unknown solution can be determined from either a calibration graph or from calculation.*

$(3.00\ ppm\ P/x\ ppm\ P) = (0.615/0.625)$

$x = (3.00)(0.625/0.615) = 3.05$ *ppm P in the solution measured*

$3.05 \times 50.0 = 152$ ppm P in the urine = 0.152 g/L

(0.152 g/L)(1.270 L/day) = 0.193 g/day

(b) (0.152 g/L)/(0.03097 g/mmol) = 4.91 mmol/L

(c) $pH = pK_2 + \log ([HPO_4^{2-}]/[H_2PO_4^-])$

$6.5 = 7.1 + \log ([HPO_4^{2-}]/[H_2PO_4^-])$

$([HPO_4^{2-}]/[H_2PO_4^-]) = 10^{-0.6} = 10^{-1} \times 10^{0.4} = 2.5 \times 10^{-1}$

43. The concentration of iron(II) in the stock standard solution is

(0.0702 g $Fe(NH_4)_2SO_4.6H_2O$/L)[(55.85 g Fe/mol)/(392.1 g Fe $(NH_4)_2SO_4.6H_2O$/mol)

= 0.0100 g/L = 10.0 mg/L = 10.0 ppm Fe

The corresponding working standards are:

Solution 1	(10.0 ppm)(1.00/100) = 0.100 ppm
Solution 2	(10.0 ppm)(2.00/100) = 0.200 ppm
Solution 3	(10.0 ppm)(5.00/100) = 0.500 ppm
Solution 4	(10.0 ppm)(10.0/100) = 1.00 ppm

A plot of A vs. ppm gives a straight line drawn through the origin. From this calibration graph and the absorbance of the sample, the sample concentration is 0.540 ppm.

(0.540 mg/L)(0.100 L) = 0.0540 mg Fe in the sample

44. The concentration of nitrate N in the standard solution is

(0.722 g KNO_3/L)[(14.07 g N/mol)/101.1 g KNO_3/mol)] = 0.100 g/L

= 100 mg/L = 100 ppm N

Hence, the amount of standard added to spiked sample is 0.100 mg, corresponding to an increase in concentration of

0.100 mg/100 mL = 1.00 mg/L = 1.00 ppm N

(The volume change can be ignored since solutions were evaporated and *ultimately diluted to a fixed volume of 50 mL.)*

The net absorbances are

Sample: $0.270 - 0.032 = 0.238$

Sample plus standard: $0.854 - 0.032 = 0.822$

The change in absorbance due to a sample concentration change of 1.00 ppm N is

$0.822 - 0.235 = 0.587$

Hence, the concentration of nitrate N in the sample is

$(1.00\ \text{ppm})(0.238/0.587) = 0.405\ \text{ppm N}$

45. *The initial concentrations of A and B are 0.00500* $\underline{M}$.

AB	$=$	A +	B
$0.00500 - x$		x	x

$(x^2)/(5.00 \times 10^{-3} - x) = 6.00 \times 10^{-4}$

Solution of the quadratic equation for x yields:

$x = 1.46 \times 10^{-3}\ \underline{M}$

Therefore, the concentration of the complex is $(5.00 - 1.46) \times 10^{-3} = 3.54 \times 10^{-3}\ \underline{M}$.

$A = 450 \times 1.00 \times 3.54 \times 10^{-3} = 1.59_3$

46. *The concentration of A can be calculated from the absorbance at 267 nm since B does not absorb there:*

$0.726 = (157\ \text{cm}^{-1}\ \text{g}^{-1}\ \text{L})(1.00\ \text{cm})\ c_A$

$c_A = 4.62 \times 10^{-3}\ \text{g/L} = 4.62\ \text{mg/L}$

$A_{312} = A_{A,312} + A_{B,312}$

$0.544 = (12.6)(1.00)(4.62 \times 10^{-3}) + (186)(1.00)c_B$

$c_B = 2.61 \times 10^{-3}$ g/L = 2.61 mg/L

47. $0.805 = k_{415,Ti} \times 1.00 \times 10^{-3}$; $k_{415,Ti} = 805$

$0.465 = k_{455,Ti} \times 1.00 \times 10^{-3}$; $k_{455,Ti} = 465$

$0.400 = k_{415,V} \times 1.00 \times 10^{-2}$; $k_{415,V} = 40.0$

$0.600 = k_{455,V} \times 1.00 \times 10^{-2}$; $k_{455,V} = 60.0$

$0.685 = 805 \times C_{Ti} + 40.0 \times C_V$

$0.513 = 465 \times C_{Ti} + 60.0 \times C_V$

$C_{Ti} = 6.93 \times 10^{-4}$ $\underline{M}$, $C_V = 3.18 \times 10^{-3}$ $\underline{M}$

6.93×10^{-4} $\underline{M}$ x 100 mL = 0.0693 mmol Ti

0.0693 mmol x 47.9 mg/mmol = 3.32 mg Ti

% Ti = (3.32 mg/1000 mg) x 100% = 0.332%

3.18×10^{-3} $\underline{M}$ x 100 mL = 0.318 mmol V

0.318 mmol x 50.9 mg/mmol = 16.2 mg V

% V = (16.2 mg/1000 mg) x 100% = 1.62% V

48. From Beer's law, fraction of energy absorbed = $1 - (P/P_o) = 1 - 10^{-abc}$

$P_o - P = P_o(1 - 10^{-abc})$ = quantity of light absorbed

$F \propto$ quantity of light absorbed

$F \propto P_o(1 - 10^{-abc})$

$F = \phi P_o (1 - 10^{-abc})$, ϕ = proportionality constant

1. Only a small fraction of atoms in a flame are thermally excited, 0.01% or less.

2. A solution is aspirated into a flame where atomic vapor of the elemental constituents (metals) is produced. A certain small fraction of the atoms is thermally excited to higher electronic levels, and when these return to the ground state, photons of characteristic wavelength are emitted. These emitted photons are measured in flame emission spectrophotometry. In atomic absorption spectrophotometry, the absorption of the characteristic wavelengths of light by the ground state atoms is measured.

3. Flame emission spectrometry requires a flame-aspirator for atomization and excitation of the sample, a monochromator for separating the emitted wavelengths, and a detector, usually a photomultiplier. Atomic absorption spectrophotometry is similar, but requires in addition a light source, usually a sharp line source. The two techniques are similar in sensitivity for a large number of elements, with each being more sensitive for a number of specific elements. They are both subject to essentially the same chemical and ionization interferences, but flame emission spectroscopy is generally more subject to spectral interferences.

4. In order to obtain higher sensitivity, greater specificity, and linearity in calibration curves.

5. Because the difference between zero and a finite number is measured, and the sensitivity is governed by the sensitivity and stability of the detector (as well as flame background noise).

6. Many atoms react with the flame gases, forming MO or MOH species and decreasing the free atom population which either technique measures. Hence, the dependence of the free atom population on temperature may be the major signal determinant and will affect either method similarly. At wavelengths above 300 nm, sufficient excitation occurs to provide good emission sensitivity.

7. Because the discrete electronic transitions are not superimposed on rotational and vibrational transitions, since atoms do not undergo these latter two modes of transition.

8. The red zone in a reducing nitrous oxide-acetylene flame results from the presence of highly reducing radical species, such as CN and NH radicals.

9. The atomization efficiency (the fraction of sample converted to atomic vapor) is much higher than in a flame, as is the residence time in the absorption path.

10. If the internal standard and analyte are chemically similar, changes in these populations due to chemical reactions, changes in aspiration rate or flame conditions and so forth will be similar for both elements. Hence, the ratio of their populations will remain constant and so will the ratio of their measured signals.

11. The cathode is constructed of the element to be determined, or an alloy of it, and the anode is an inert metal such as tungsten. The lamp is filled with an inert gas under reduced pressure. A high voltage across the electrodes causes the gas to ionize at the anode, and these positive ions bombard the cathode, causing the metal to sputter (vaporize) and become electronically excited. These excited atoms emit the characteristic lines of the metal when they return to the ground state.

12. In the premix chamber burner, the fuel and support gases are introduced into a chamber where they are mixed before entering the burner head where they combust. It is limited to relatively low-burning velocity flames, such as air-acetylene. It cannot be used with oxygen-supported flames.

13. To convert it to an a.c. signal which can be selectively measured with an a.c. detector tuned to its modulation frequency and thereby discriminate from d.c. emission from the flame.

14. Probably non-specific background absorption by something in the sample extract. It can be corrected for by measuring the absorbance of a non-absorbing (for lead) line a few nanometers away (e.g., the thallium 276.8 nm line) and subtracting from the sample signals.

15. In order to atomize refractory elements like Al and V, that form stable oxides in cooler flames.

16. To suppress ionization.

17. To minimize ionization of the elements. They have few chemical interferences.

18. The presence of the potassium suppresses ionization of the sodium, causing an increase in free sodium atoms and hence atomic emission.

19. Sensitivity = concentration needed to give 1% absorption or 0.0044 A

(= -log 0.99)

$T = 0.920$

$A = \log 1/0.920 = 1.09 = 0.0374$

(0.0044 A/0.0374 A) = (x ppm/12 ppm)

x = 1.4 ppm per 0.0044 A = sensitivity

20. (0.050 ppm/0.70 ppm) = (0.0044 A/x A); x = 0.062 A

-log T = 0.062; T = 0.87; % absorbed = 13%

21. T = 2250° C = 2523 K

From the term symbols, $J_e = 1$ and $J_o = 0$

$g = 2J + 1$

$g_e/g_o = [2(1) + 1]/[2(0) + 1] = 3/1$

$\lambda = 228.8 \text{ nm} = 2.288 \times 10^{-5} \text{ cm}$

$\nu = c/\lambda = (3.00 \times 10^{10} \text{ cm g}^{-1})/(2.288 \times 10^{-5} \text{ cm}) = 1.31 \times 10^{15} \text{ s}^{-1}$

$E_e - E_o = h\nu = (6.62 \times 10^{-27} \text{ erg-s})(1.31 \times 10^{15} \text{ s}^{-1}) = 8.68 \times 10^{-12} \text{ erg}$

$N_e/N_o = g_e/g_o \, e^{-h\nu/kT}$

$= (3/1) \exp [-(8.68 \times 10^{-12} \text{ erg})/(1.380 \times 10^{-16} \text{ erg K}^{-1})(2523 \text{ K})]$

$= 3e^{-24.93} = 4.5 \times 10^{-11}$

$\% N_e = [N_e/(N_e+N_o)] \times 100\% = [(4.4 \times 10^{-11})/(1.00)] \times 100\% = 4.5 \times 10^{-9}\%$

22. Stock Ca solution:

$1.834\ g\ [(40.08\ g/mol\ Ca)/147.02\ g/mol\ CaCl_2.2H_2O] = 0.500\ g\ Ca$

500 mg Ca/L = 500 ppm Ca

Therefore, second stock solution = 50.0 ppm

Working standards = 2.50, 5.00, and 10.0 ppm Ca

Net signals (subtract 1.5 cm for blank):

standards: 9.1, 18.6, 37.0 cm

sample: 28.1 cm

Plot of net signal vs. concentration of standards gives straight line through origin. Sample reading on the graph corresponds to 7.62 ppm Ca. Therefore, concentration in original sample is

7.62 ppm x 25 = 190 ppm Ca

23. The volume of added $LiNO_3$ can be neglected compared to 1 mL (1% change - the instrumental reproducibility is about 2%). Hence, the concentration of Li added to the sample solution is 0.010 $\underline{M}$ x (0.01/1) = 1.0×10^{-4} $\underline{M}$. The net increase in signal is due to this added concentration: 14.6 - 6.7 = 7.9 cm. Assuming linearity, a direct proportionality applies:

$(7.9\ cm)/(1.0 \times 10^{-4}\ \underline{M}) = (6.7\ cm)/(x)$

$x = 8.5 \times 10^{-5}$ $\underline{M}$ in the diluted serum

$8.5 \times 10^{-5}\ \underline{M} \times (1/0.1) = 8.5 \times 10^{-4}$ $\underline{M}$ in the undiluted serum

$8.5 \times 10^{-4}\ mmol/mL \times 6.94\ mg/mmol = 5.9 \times 10^{-3}\ mg/mL$

$= 5.9\ \mu g/mL = 5.9\ ppm$

24. A direct proportionality exists between concentration and the decrease in the silver absorbance signal from the blank reading.

$(12.8 - 5.7\ cm)/(100\ ppm) = (12.8 - 6.8\ cm)/(x)$

x = 84 ppm in sample

1. The distribution coefficient is the ratio of the concentration of the extracted solute in the nonaqueous phase to its concentration in the aqueous phase. The distribution ratio is the ratio of the concentration of all forms of the solute in the two phases and accounts for ionization, dimerization, etc.

2. Extract the nitrobenzene from acid solution into benzene. The aniline is ionized ($C_6H_5NH_3^+$) in the acid and will not extract.

3. (1) Ion-association extraction systems: The metal ion is incorporated into a bulky molecule, either charged or uncharged, which associates with another ion of the opposite charge to form an ion-pair, or else the metal ion associates with another ion of great size. An example is the coordination of the chloro complex of iron(III) with ether and association with the protonated solvent, i.e., the solvent extraction of iron(III) from HCl solutions into ether.

(2) Metal chelate extraction systems: The metal ion forms an uncharged chelate with an organic chelating agent which is soluble in the nonaqueous solvent. An example is the solvent extraction of aluminum with 8-hydroxyquinoline into chloroform.

4. The chelating agent, a weak acid, distributes between the two phases, it ionizes in the aqueous phase, the ionized reagent chelates with the metal ion, and the chelate distributes between the two phases.

5. Theoretically, the upper limit is the solubility limit of the chelate in the organic solvent, and at the other extreme, unmeasurable amounts can be extracted.

6. An increase in pH by 1 unit or an increase in reagent concentration by ten-fold will increase the extraction efficiency the same amount.

The distribution ratio is proportional to $[HR]^n/[H^+]^n$, where n is the number of reagent molecules chelated by the metal ion.

7. The sample and solvent are heated in a closed vessel under pressure to temperatures above the ambient boiling point of the solvent. The elevated temperature accelerates dissolution of analytes in the solvent.

8. The sample and solvent are heated in a microwave oven in a closed vessel, which accelerates the extraction process.

9. Extractions are achieved using microparticles coated with hydrophobic functional groups. These are the same particles used in high performance liquid chromatography. The hydrophobic layer acts as the extracting solvent. The particles are placed in small cartridges, pipet tips, or on disks. Many samples can be processed simultaneously and automatically. Small volumes of eluting solvent are needed, reducing organic solvent waste.

10. SPME is a solvent-less extraction technique in which analytes are adsorbed onto a micro fused silica fiber coated with a solid adsorbent or an immobilized polymer. Following adsorption, the analyte is thermally desorbed.

11. Equation 18.9:

$$\% E = \{([S]_o V_o)/([S]_o V_o + [S]_a V_a)\} \times 100 \quad (1)$$

Also,

$$D = ([S]_o/[S]_a) \text{ (assuming no side reactions)} \quad (2)$$

Divide the numerator and denominator of (1) by $[S]_a$:

$$\% E = \{[([S]_o/[S]_a)V_o]/[([S]_o/[S]_a)V_o + V_a]\} \times 100 = [(DV_o/(DV_o + V_a)] \times 100$$

Divide top and bottom by V_o:

$$\% E = [D/(D + V_a/V_o)] \times 100$$

12. The concentration of benzoic acid present in the dimer is twice the dimer concentration. Hence,

$$D = ([HBz]_e + 2[(HBz)_2]_e)/([HBz]_a + [Bz^-]_a) \quad (1)$$

From Equation 16.6,

$$[HBz]_e = K_D[HBz]_a \quad (2)$$

For the dimer, $[(HBz)_2]_e = K_p[HBz]_e^2 = K_pK_D^2[HBz]_a^2$ (3)

From Equation 18.5,

$$[Bz^-]_a = K_a[HBz]_a/[H^+]_a \quad (4)$$

Substituting (2), (3), and (4) in (1),

$$D = (K_D[HBz]_a + 2K_pK_D^2[HBz]_a^2)/([HBz]_a + K_a[HBz]_a/[H^+]_a)$$

$$D = K_D(1 + 2K_pK_D[HBz]_a)/(1 + K_a/[H^+]_a)$$

This is identical to Equation 18.8 except the term in parentheses in the numerator is added. From this, the distribution ratio and extraction efficiency are now dependent on the concentration of the extracting species, in addition to the pH.

13. x = % extracted in a single step

$x + (100 - x)(x/100) = 96$

x = 80% extracted in a single extraction

$80\% = (100\ D)/(D + 100/50)$

$D = 8.0$

14. $\%\ E = (100 \times 2.3)/(2.3 + 25.0/10.0) = 48\%$

15. At $V_a = V_o$, $90\% = (100\ D)/(D + 1)$

$D = 9$

∴ when $V_a = 0.5\ V_o$,

$\%\ E = (100 \times 9)/(9 + 0.5) = 95\%$

16. With $V_o = V_a = 10$ mL,

$\%\ E = (100 \times 25.0)/(25.0 + 1.0) = 96.2\%$ extracted, leaving 3.8%.

With $V_o = 0.50\ V_a = 5.0$ mL,

$\%\ E\ (100 \times 25.0)/(25.0 + 2.0) = 92.6\%$ extracted first time, leaving 7.4%.

On second extraction,

$0.074 \times 7.4 = 0.55\%$ remains.

∴ Two extractions with half the volume of organic solvent is more efficient.

17. 1st extraction: 30% remains

2nd extraction: $0.30 \times 30 = 9.0\%$ remains

3rd extraction: $0.30 \times 9.0 = 2.7\%$ remains

1. Chromatography is a physical method of separation in which the components to be separated are distributed between two phases, one of which is stationary, while the other moves in a definite direction.

2. There is an equilibrium distribution of the solutes between two phases, one mobile and one stationary. The solutes are eluted from the stationary phase by movement of the mobile phase. Due to differences in the distribution equilibria for different solutes, they are eluted at different rates.

3. Chromatographic techniques include adsorpton, partition, ion exchange, and size exclusion. Gas chromatography utilizes the first two, while liquid chromatography utilizes all four.

4. The van Deempter equation describes H as a function of velocity of carrier gas or eluent: $H = A + B/\bar{u} + C\bar{u}$. A is the eddy diffusion term, B the molecular diffusion term, and C the mass transfer term.

5. The Golay equation applies to open tubular columns. It does not contain the eddy diffusion (A) term since there is no packing.

6. The Huber and Knox equations contain terms that account for mass transfer in both the stationary and the mobile phase.

7. $N = 16\ (65/5.5)^2 = 2.2_3 \times 10^3$ plates

$$H = (3\text{ ft} \times 12\text{ in/ft} \times 2.54\text{ cm/in})/(2.2_3 \times 10^3\text{ plates}) = 0.041\text{ cm/plate}$$

8. When the two peaks are just resolved, the peak base widths will be 15 s, since the difference between their retention times is 15 s. The number of theoretical plates required to elute the last peak is then:

$$N = 16\ (100/15)^2 = 710\text{ plates}$$

The column length is:

$$710\text{ plates} \times 1.5\text{ cm/plate} = 1.0_6 \times 10^3\text{ cm}$$

9. $N_{eff} = 16\ (t_R'/w_b)^2$; $H_{eff} = L/N = 300$ cm/N (Use of t_R' gives the effective theoretical plates)

mL/min	t_R'	t_R'/w_b	N	H (cm/plate)$_{eff}$
120.2	4.31	12.3	2420	0.123
90.3	4.88	12.5	2500	0.120
71.8	5.43	12.6	2550	0.118
62.7	5.73	12.2	2380	0.126
50.2	6.38	11.8	2230	0.135
39.9	7.25	10.7	1830	0.164
31.7	8.21	10.1	1640	0.183

The optimum flow velocity is near 75 mL/mm.

10. Calculate the α-values between the peaks, and from these the number of plates needed to obtain the more difficult one.

$$\alpha_{AB} = 1.85/1.40 = 1.32$$
$$\alpha_{BC} = 2.65/1.85 = 1.43$$

Peaks A and B are the poorest resolved, and so a separation factor of 1.32 is needed. The mean value of the two peaks is (1.40+1.85)/2 = 1.62. From Equation 19.30, the number of plates required for the separation is:

$$N_{req} = 16(1.05)^2[1.32/(1.32+1)]^2[(1.62+1)/1.85]^2 = 602 \text{ plates required.}$$

Hence, compounds A and B won't quite be separated with a resolution of 1.05. The actual resolution of the two peaks is:

$$R_s = \tfrac{1}{4}(500)^{1/2}[(1.32-1)/1.32][1.85/(1.62+1)] = 0.96$$

11. See CD for spreadsheet and chart of van Deempter plot.

CHAPTER 20

1. *The stationary phase is a solid or liquid, and the mobile phase is a gas. The solute exists in the vapor state, usually accomplished by heating, and it distributes between the stationary phase and the gas phase.*

2. *All gases, most non-ionized organic molecules up to C25, volatile derivatives of non-volatile compounds.*

3. *Gas-solid and gas-liquid are the two types of gas chromatography.*

4. *Packed columns can have 1,000 plates/m, and a representative 3 m column has 3,000 plates. a capillary column (WCOT) typically has 5,000 plate/m, and 250,000 plates for a 50 m column.*

5. *Wall coated open-tubular (WCOT) columns have thin film of stationary phase on the capillary wall. Support coated open-tubular (SCOT) columns have solid microparticles coated with stationary phase attached to the wall. They have higher capacity but lower resolution. Porous layer open-tubular (PLOT) columns have solid particle attached to the wall for adsorption chromatography.*

6. *(a) Thermal conductivity. The difference in the thermal conductivity of the carrier gas and the carrier gas plus solute is measured with a Wheatstone bridge, as the solute is eluted. Solutes generally have a lower thermal conductivity than the carrier gas. A carrier gas with a high thermal conductivity is used, e.g., helium or hydrogen.*

(b) Flame ionization. The eluted solutes are burned in a hydrogen flame to produce ions which are collected by a pair of charged electrodes. The resulting current is measured. This detector is insensitive to water, allowing separation of solutes in aqueous solution.

(c) Electron capture. An electron source (electrical or b-ray) provides a preselected current to an anode collector. Compounds which capture electrons cause a decrease in this current, which is recorded. A carrier gas with a low excitation energy is used, such as nitrogen or hydrogen, to prevent ionization of compounds.

7. *(a) Thermal conductivity. A general detector with relatively low sensitivity.*

(b) Flame ionization. A general detector with high sensitivity, about 1000 times that of the T. C. detector.

(c) Electron capture. Specific for a limited number of substances with high electron capture affinity, e.g., chlorine-containing compounds. High sensitivity.

8. *A lower initial temperature can be used so faster eluting compounds can be better resolved, and higher temperatures for more strongly retained compounds so they elute faster with less broadening.*

9. *For fast GC, use a narrow, short column with thin stationary film and a light carrier gas. A fast responding detector is required.*

10. The detector is a mass spectrometer, which creates ions and fragments that are separated according to their mass-to-charge ratios and then detected by an electron multiplier. Mass spectra from individual chromatographic peaks can be recorded, providing definitive identification via the fragmentation pattern and/or detection of the parent ion peak (giving the molecular weight). The method is capable of high sensitivity.

11. The molecular ion in the M^+, where M is the molecular weight of the analyte molecule. It may or may not appear in the mass spectrum. The base peak is the largest one, and others are normalized to it at 100% relative abundance.

12. A compound with no or an even number of nitrogen atoms will have an even molecular mass. Otherwise it will be odd.

13. The electron-impact (EI) and the chemical ionization (CII) ionization sources are commonly used for GC-MS. CI gives the molecular ion whereas EI may not.

14. The quadrupole mass filter is the most commonly used analyzer for GC-MS. The time-of-flight (TOF) is the next most common.

15. $R = m/\Delta m = 600/1 = 600$

16. $m/\Delta m = R$

$600/\Delta m = 5{,}000$

$\Delta m = 0.12$

So they are resolved at 0.12 mass unit, i.e., 600.00 is resolved from 600.12.

17. 760 torr/14.7 psi = 51.7 torr/psi

40.0 psig x 51.7 torr/psi = 2070 torr above atmospheric pressure

2070 torr + 740 torr (atm. press.) = 2810 torr inlet pressure

18. See CD. y = 0.0697x – 0.0316. The calculated concentration is 20.9 ppm.

1. High performance liquid chromatography uses smaller, more uniform particles with thin stationary phases to allow more rapid mass transfer (diffusion) processes (small C in van Deemter equation). The result is that more rapid equilibrium is achieved, allowing rapid flow rate (which also decreases molecular diffusion - B in van Deemter equation). The small particles create a large pressure drop, however, and the eluent must be pumped through the column at high pressures. Small columns are used so that microgram or smaller quantities of solutes can be separated. Automatic detectors are also used to obtain a direct and rapid recording of the chromatogram.

2. The two most commonly used detectors are the differential refractometer and the ultraviolet detectors. The former measures the difference in the index of refraction of the solvent and the solvent plus solute as the solute is eluted. The latter measures the absorbance by solutes in the ultraviolet region as they are eluted.

3. They would elute in decreasing order of polarity, i.e., CH_3CHO, CH_3CH_2OH, CH_3CO_2H.

4. A nonpolar solvent like hexane or heptane.

5. In normal phase chromatography, a polar stationary phase is used, with a nonpolar mobile phase. It is used for separating polar compounds. Reversed-phase chromatography uses a nonpolar stationary phase and polar mobile phase, and is used for separating polar compounds.

6. Reversed-phase HPLC is the most commonly used because it is useful for separating a wide range of organic compounds.

7. Silica particles are endcapped to bond the residual silanol groups and render them inert.

8. For reversed-phase HPLC: those with C_{18}, C_8, C_5, and phenyl groups. For normal phase HPLC: those with cyano, diol, amino, and dimethylamino groups.

9. Microporous particles have small pores, and the solute must diffuse in and out of these. Perfusion particles have a mixture of large and small pores, allowing the solution to flow directly through the particles, increasing mass transfer. They can be used at higher flow rates and have better efficiencies for biomolecules. Nonporous particles have a solid core, with a thin porous layer. They are smaller and eliminate stagnant mobile phase, giving increased mass transfer, but at the expense of higher backpressure and limited capacity. They are used in ion chromatography.

10. A guard column is a small column with usually the same packing as the analytical column, placed between the injector and the column. It retains debris and sample

particulate matter, and highly sorbed sample matrix compounds. It extends the life of the analytical column.

11. The multi-solvent method is a trial-and-error procedure of varying the solvent strength to achieve optimum retention (k) and separation (α).

12. Gradient elution is used to separate analytes with substantial differences in retention. A weaker solvent is used at the start so weakly retained compounds can be resolved, and the fraction of organic modifier (in RPC) is increased, either stepwise or gradually, to elute the more strongly retained compounds faster and with sharper peaks.

13. In fast LC, columns are packed with smaller particles (1-3 μm instead of 5 μm) to increase mass transfer, and they are shorter (1-3 cm instead of 15-25 cm) because of increased backpressure. A fast detector is needed.

14. Narrowbore (2.1 cm) columns give narrower, taller peaks and increased sensitivity, about five-fold. Or a five-fold smaller sample can be injected with the same sensitivity.

15. Elevated temperature does not affect separation efficiencies much in HPLC, in contrast to in GC, and mobile phase modification is more effective. But in fast separations using high mobile phase velocity, high temperature does increase efficiency for strongly retained solutes.

16. The common interfaces in LC-MS are the electrospray ionization source (EI), the thermospray ionization source (TI), the atmospheric pressure chemical ionization source (APCI), and the particle beam ionization source. ESI is preferred for polar, ionic, and very large molecules. APCI is complementary toe EI in that less polar analytes are more efficiently ionized, but it is limited to 2000 daltons. Thermospray is a soft ionization technique, and is useful for non-volatile, thermally labile organic compounds, although it is largely replaced now by APCI. The particle beam gives spectra similar to EI spectra obtained in GC-MS, making interpretation easy.

17. The stationary phase is a molecular sieve that has an open network, which excludes solutes above a certain size. Solutes of formula weight above the limit are not retained by the column and can be separated from smaller solutes that are retained. The exclusion limit is the formula weight of the molecule that is excluded (not retained) by the molecular sieve.

18. A polymeric carbohydrate or acrylamide with an open network formed by the cross-linking of the polymeric chains.

19. A cation exchange resin contains acidic functional groups, the anion of which is capable of associating with and exchanging cations. An anion exchange resin contains basic functional groups, the cation of which is capable of associating with and exchanging anions.

20. Charge, ion size, degree of cross-linkage of the resin, temperature, solvent, pH.

21. In ion chromatography, the eluted analyte and the eluent emerging from an analytical ion exchange column are passed through a high capacity suppressor column (cation exchanger, H^+ form for base eluent; anion exchanger, OH^- form for acid eluent) to convert the eluent to H_2O or a low conducting acid. This allows sensitive continuous conductometric detection of the analytes as they emerge separately, to provide an ion chromatogram.

22. The retention time is the time required for a solute to elute from a chromatographic column under specified conditions, usually in gas chromatography. It is measured from the time zero to the the time corresponding to the peak of the chromatogram of the solute. The R_f value is the ratio of the distance the solute moves to the distance the solvent front moves in paper or thin layer chromatography. Both of these values are qualitative measures of the solute.

23. Largest value unity, smallest value zero

24. Separation is based on chromatographic equilibria plus interaction of charged solutes with an electric field.

25. The sample travels through a capillary by means of electroosmosis, by inserting the ends of the capillary in a buffer solution and applying a high D.C. voltage by means of platinum electrodes immersed in the solution. Uncharged analytes migrate together at the electroosmotic rate, but charged analytes travel at different rates based on their electrophoretic mobilities. The method is capable of very high resolution (10^6 theoretical plates) and sensitivity (10^{-21} mol), allowing rapid separation of high molecular weight molecules.

26. mmol K^+ = mmol H^+ = mmol NaOH

mmol K^+ = 0.0506 mmol/mL x 26.7 mL = 1.35 mmol

1.35 mmol/5.00 mL = 270 mmol/L

27. meq = mmol for a monovalent ion

(10 g/L)/(58 g/mol) = 0.17 mol/L = 170 mmol/L

170 mmol/L x 0.20 L = 34 mmol Na^+ = 34 meq

(34 meq)/(5.1 meq/g) = 6.7 g resin

28. (a) HCl (b) H_2SO_4 (c) $HClO_4$ (d) H_2SO_4

CHAPTER 22

1. *The rate of a first-order reaction is proportional to the concentration of a single reactant. The rate of a second-order reaction is typically proportional to the product of the concentrations of two reactants. (Strictly speaking, the order of a reaction is the sum of the exponents of the concentration terms in a rate equation.)*

2. *The half-life is the time it takes for a reaction to go to 50% completion. Theoretically, it takes an infinite number of half-lives to go to 100% completion, but practically speaking, a reaction can be considered essentially complete in about 10 half-lives.*

3. *A pseudo first-order reaction is a higher order reaction whose rate is made to depend on the concentration of a single reactant by making the concentrations of the other reacting species high enough that they become essentially constant.*

4. *Set the starting concentrations of each equal. Plot log [A] against t (Equation 22.3). If the reaction is first order, a straight line will result, and if it is not, a curved line will result. Plot $([A]_0 - [A])/[A]$ vs. t (Equation 22.7). If the reaction is second order, a straight line will result.*

5. *An international unit (I.U.) is the amount of enzyme that will catalyze the transformation of one micromole of substrate per minute under defined conditions.*

6. *A competitive inhibitor competes with the substrate for an active site on the enzyme, and the inhibition varies with the concentration of the substrate. A noncompetitive inhibitor combines with the enzyme at a site other than the active site to form an inactive derivative of the enzyme, and the inhibition is independent of the concentration of the substrate.*

7. *They frequently combine with sulfhydryl groups in enzymes to inactivate the enzymes.*

8. *Substances which activate enzymes.*

9. *Double the substrate concentration and measure the rate of the reaction. If the inhibition is competitive, the percent inhibition will decrease, whereas if it is noncompetitive, it will remain the same.*

10. *If the inhibition is competitive, the slope of the plot will change, but the intercept will be the same. If the inhibition is noncompetitive, the intercept will change.*

11. $t_{1/2} = (0.693/k)$

$10.0\ min = (0.693/k)$

$k = 0.0693\ min^{-1}$

$\log [A] = \log [A]_0 - (kt/2.303)$

Let $[A]_0 = 1$, then $[A]$ at 90% conversion = 0.1

$\log 10^{-1} = \log 1 - (0.0693\ t/2.303)$

$t = 33.2\ min$

At 99% conversion, $[A] = 0.01$

$\log 10^{-2} = \log 1 - (0.0693\ t/2.303)$

$t = 66.5\ min$

12. Let $[A]_0 = 1$, then $[A] = 0.7$ at 25.0 s

$\log 0.7 = \log 1 - (k \times 25.0)/(2.303)$

$k = 0.0143\ s^{-1}$

$t_{1/2} = (0.693/0.0143) = 48.5\ s$

13. $kt = ([A]_0 - [A])/([A]_0[A])$

$k \times 6.75 = [(0.100 - 0.850 \times 0.100)]/[0.100 \times (0.850 \times 0.100)]$

$k = 0.26\ min^{-1}\ \underline{M}^{-1}$

$t_{1/2} = 1/(k[A]_0) = 1/(0.26\ min^{-1}\ \underline{M}^{-1} \times 0.100\ \underline{M}) = 38\ min$

At 0.200 $\underline{M}$:

$t_{1/2} = 1/(0.26 \times 0.200) = 19\ min$

$0.26\ t = [(0.200 - 0.850 \times 0.200)]/[0.200 \times (0.850 \times 0.200)]$

$t = 3.4\ min$ for 15.0% completion

14. At 25.0% conversion, [sucrose] = $0.750 \times 0.500\ \underline{M} = 0.375\ \underline{M}$

$\therefore$ log 0.667 = log 1 - (0.0320 x t)/(2.303)

t = 12.7 h

15. At 35%, the fraction of H_2O_2 remaining is 0.650.

$kt = ([A]_0 - [A])/([A]_0[A])$

$([A]_0 = [B]_0 = 1/2[H_2O_2]_0)$

k x 8.60 = [(0.0500 - 0.650 x 0.0500)]/[0.0500 x (0.650 x 0.0500)]

$k = 1.25\ min^{-1}\ \underline{M}^{-1}$

100 mL of O_2 at STP = (100 mL)/(22,400 mL/mol) = 0.00446 mol = 4.46 mmol O_2

Twice this many millimoles of H_2O_2 are used in forming the O_2 = 8.92 mmol

mmol H_2O_2 at start = 0.1000 $\underline{M}$ x 100.0 mL = 10.00 mmol

mmol H_2O_2 remaining = 10.00 - 8.92 = 1.08 mmol in 100 mL = 0.0108 $\underline{M}$

$\therefore$ 1.25 t = [(0.0500 - 1/2 x 0.0108)]/[0.0500 x (1/2 x 0.0108)]

t = 132 h

16. µmol O_2 consumed/20 min = (10.5 mL)/(0.0224 mL/µmol) = 469 µmol/20.0 min

= 23.4 µmol/min

1 unit transforms 1 µmol/min

$\therefore$ activity of enzyme preparation = 23.4 units/10.0 mg = 2.34 units/mg

% purity = (2.34/61.3) x 100% = 3.82%

17. See CD for spreadsheet and chart of Lineweaver-Burk plot.
Slope = 11.9 ΔA^{-1}min·mM
Intercept = $1/R_{max}$ = 9.97 ΔA^{-1}min
R_{max} = 0.100 $\Delta A min^{-1}$
K_m = slope x R_{max} =1.19 mM

1. An automatic instrument performs one or more steps of an analysis without human instruction. An automated instrument performs all steps of an analysis and uses the information to regulate a process without human intervention.

2. A continuous automated instrument continuously and constantly makes measurements on a process stream and feeds this information to controlling devices to continuously regulate the process. A discrete automated device makes measurements at discrete intervals to provide information to the regulator in discrete steps. The controlled variable is maintained at a fixed level between measurements.

3. A discrete sampling instrument analyzes each sample in a separate cuvet or chamber. A continuous flow sampling instrument analyzes the samples sequentially in a single tube. They are usually separated by air bubbles.

4. A feedback control loop utilizes measured information from a sensor about a variable to be controlled to compare against a set point in a controller. The controller feeds any difference to an operator which activates a regulating device to bring the variable back to the set point.

5. Flow injection analysis (FIA) is an unsegmented continuous flow technique in which a few microliters of sample are injected into a flowing reagent stream. Mixing occurs by diffusion and in about 15 s, product is detected as it flows through a micro-flow cell detector, resulting in a peak whose height is proportional to the sample concentration.

6. SIA is a single-line, single reversible pump system that uses a multiposition selection valve for aspirating reagent and sample plugs into a holding coil, then propels them to the detector. The adjacent zones merge and react on the way to the detector. It differs from FIA in that only a few microliters of reagent are used, and chemistries are readily changed by selecting a different reagent port, using the computer.

7. Computers are used to speed up certain parts of the analytical process. This includes acquiring and processing data, sometimes large amounts, and controlling the instrumentation and experiments. Several instruments on experiments can be controlled simultaneously.

8. Microprocessors can perform automatic background correction, take the derivative of a spectrum, integrate or average signals, and automatically calculate concentrations and standard deviations.

CHAPTER 23

9. (a) From Equation (22.3),

$1 - 1/D = 10^{-kS_V}$ or $\log(1 - 1/D) = -kS_V$

$\log(1 - 1/D_1)/\log(1 - 1/D_2) = S_{V1}/S_{V2}$

$\log(1 - 1/4.00)/\log(1 - 1/D_2) = 1/2$

$-0.250 = \log(1 - 1/D_2)$

$1 - 1/D_2 = 10^{-.25} = 10^{-1} \times 10^{.75} = 0.562$

$D_2 = 2.28$

D decreased by a factor of 2.28/4.00 = 0.570 on doubling the sample volume, while in Example 19.2, it decreased by 1.80/3.01 = 0.598, about a 5% difference. That is, the larger D (the more the dilution), the more it will increase on making the volume smaller.

(b) $D \propto R^2$

Therefore, doubling the tube diameter would increase D by four-fold to 16.0.

(c) $D \propto L^{1/2}$

Therefore, doubling the length would increase D by a multiple of $\sqrt{2} = 1.414$ or $D = 5.66$.

10. $1/D = 1 - 2S\ /S_{1/2}$

$1/4.00 = 1 - 2(50.0\ \mu L/S_{1/2})$

$S_{1/2} = 133\ \mu L$

1. Serum, fibrinogen, and cells. Plasma contains serum and fibrinogen, and the cells include erythrocytes, leukocytes and platelets. When blood clots, the fibrinogen precipitates, removing the cells with it and leaving serum. When unclotted whole blood is centrifuged, the cells are separated from the plasma. Blood contains dozens of chemical constituents.

2. Hemolysis is the destruction of red cells with the subsequent release of cellular constituents into the plasma or serum. The concentrations of a number of cellular constituents are higher than in the plasma and this would lead to erroneous results, if the plasma or serum is analyzed.

3. It acts as a glycolytic enzyme inhibitor to prevent the breakdown of glucose.

4. Because the red cells are ruptured, releasing cellular constituents into the plasma.

5. Blood glucose and urea (BUN) analysis.

6. An antibody is a high molecular weight immunoglobulin that specifically reacts with an antigen (foreign body).

7. The labels used in immunoassays include radiolabels, fluorescent labels,and enzyme labels. The immunoassay may be homogeneous or heterogeneous (requiring separation). Assays may be competitive or noncompetitive.

CHAPTER 25

1. Chromosomes are made up of thousands of genes. There are 23 chromosomes pairs in the nucleus of each cell (except sperm and egg cells).

2. DNA is a helical polymer consisting of nucleotides with four different nucleic acid bases, adenine (A), cytosine (C), guanine (G), and thymine (T). The bases A and T, and G and C pair by hydrogen bonding to form the "steps" of the helical DNA.

3. The polymerase chain reaction (PCR) is a technique for replicating trace quantities of DNA. The DNA is heated and cooled through several cycles in the presence of the four nucleotides, a primer, and a DNA polymerase. The heating causes the DNA strands to separate (denature). The polymerase catalyzes extension of the primer to produce a complement strand of the DNA template, thereby doubling the number of DNA molecules in each cycle.

4. Plasmids and bacterial artificial chromosomes (BACs) are vectors in which DNA fragments are inserted. These are put in bacteria where they are replicated. Plasmids insert relatively small DNA fragments (2-20 kb), while BACs insert large fragments (100-400 kb).

5. A genomic library is a set of BAC replicate DNA fragments, from a partially digested genome. These are provided to DNA sequencing laboratories, which further digest them, replicate the smaller fragments, and then determine the base sequence in the overlapping fragments.

6. BAC clones of up to 300 kb are further fragmented using nucleases, to give strands of 100-300 bases. These are replicated by insertion into plasmids. The plasmid clones are subjected to PCR replication in the presence of dideoxynucleotides,, which results in complementary strands of every length, terminated by one of four dideoxynucleotides. These are separated in order of length using electrophoresis, and identified by the color of the end nucleotide fluorescence. Overlapping fragments are aligned to identify the entire DNA sequence.

7. A SNP is a single nucleotide polymorphism. These represent about 0.1% of the genes that result in the features that make us all different, and contribute to disease and dysfunction. These genes differ by a single nucleotide.

8. DNA chips are known single strand DNA fragments that are tethered to a plate, in the form of a microarray. Unknown DNA fragments samples are identified from the fluorescent spots created when the unknown binds to its complement tether.

9. Expression profiling is the study of gene expression by comparing mRNA sequences between known and unknown samples, for example, comparing a patient's expression with that of a known disease. DNA (RNA) chips are used to identify the mRNA sequences.

10. Genomics is the study of the DNA expression (encoding) process, to form proteins in the cell. Proteomics is the study of the complements of protein structure and function in a cell.

11. Proteins consist of a polymeric sequence of amino acids, derived from twenty different amino acids.

12. The gene DNA is transcribed into an m-RNA in the cell nucleus, which migrates into the cytoplasm where it serves as the template for protein production; 3-base codons encode for a particular amino acid.

13. 2-D PAGE is 2-dimensional polyacrylamide gel electrophoresis. It separates proteins, first by isoelectric focusing based on charge, and then by gel electrophoresis based on molecular size.

14. MALDI-TOF is matrix assisted laser desorption ionization – time-of-flight mass spectrometry. It is a soft ionization technique for large molecules that produces single peaks from singly charged molecules. It is used to identify the masses of peptides in protein digestions mixtures.

15. Theoretical protein amino acid fingerprints are constructed from a combination of a knowledge of the gene codons and a large protein database. The experimental fingerprint is compared with the theoretical ones to identify the protein from fingerprints of as few as 5 or 6 peptides.

16. The first sequence is:

*GATCCA**ATTGCAT***

The second is:

*GATCCA**CATTCCGTA***

They overlap as follows:

```
GATCCAATTGCAT
      GATCCACATTCCGTA
```

The sequence is:

ATTCGATTCCGTA

1. Vacuum source, meter, collector

2. Impingers are used to sample aerosols in air.

3. Acidity, alkalinity, BOD, DO, conductivity, CO_2, Cl_2, F^-, NH_3, PO_4^{3-}, NO_3^-, SO_3^{2-}, metal ions, etc.

4. Protect from heat and light.

5. pH, dissolved gases, temperature

6. For 1 mole, V = (1 mol)(0.082 L atm/K mol)(293 K)/(1 atm) = 24.0 L

$(2.8 \times 10^{-6}\ L/L_{air})/(24.0\ L/mol) = 1.1_7 \times 10^{-7}\ mol/L_{air}$

$(1.1_7 \times 10^{-7}\ mol/L_{air})(28.0\ g/mol) = 3.3 \times 10^{-6}\ g/L_{air}$

7. 0.50 μg/L = 0.50 ppb = 500 ppt

(97/128)x500 ppt = 380 ppt toluene.

6711871R0

Made in the USA
Lexington, KY
13 September 2010